Macmillan Building and Surveying Series

Series Editor: Ivor H. Seeley
Emeritus Professor, The Nottingham Trent University

Accounting and Finance for Buildi[illegible]
A.[illegible]
Adv[illegible] second edition Ivor H. Seeley
Advanced Valuation Diane Butler and David Richmond
Applied Valuation, second edition Diane Butler
Asset Valuation Michael Rayner
Building Economics, fourth edition Ivor H. Seeley
Building Maintenance, second edition Ivor H. Seeley
Building Maintenance Technology Lee How Son and George C.S. Yuen
Building Procurement Alan E. Turner
Building Project Appraisal Keith Hutchinson
Building Quantities Explained, fourth edition Ivor H. Seeley
Building Surveys, Reports and Dilapidations Ivor H. Seeley
Building Technology, fifth edition Ivor H. Seeley
Civil Engineering Contract Administration and Control, second edition Ivor H. Seeley
Civil Engineering Quantities, fifth edition Ivor H. Seeley
Civil Engineering Specification, second edition Ivor H. Seeley
Commercial Lease Renewals – A Practical Guide Philip Freedman and Eric F. Shapiro
Computers and Quantity Surveyors A.J. Smith
Conflicts in Construction – Avoiding, managing, resolving Jeff Whitfield
Constructability in Building and Engineering Projects A. Griffith and A.C. Sidwell
Construction Contract Claims Reg Thomas
Construction Economics – An Introduction Stephen L. Gruneberg
Construction Law Michael F. James
Contract Planning and Contractual Procedures, third edition B. Cooke
Contract Planning Case Studies B. Cooke
Cost Estimation of Structures in Commercial Buildings Surinder Singh
Design–Build Explained D.E.L. Janssens
Development Site Evaluation N.P. Taylor
Environmental Management in Construction Alan Griffith
Environmental Science in Building, third edition R. McMullan
Estimating, Tendering and Bidding for Construction A.J. Smith
European Construction – Procedures and techniques B. Cooke and G. Walker
Facilities Management Alan Park
Greener Buildings – Environmental impact of property Stuart Johnson
Housing Associations Helen Cope
Housing Management – Changing Practice Christine Davies (Editor)
Information and Technology Applications in Commercial Property Rosemary Feenan and Tim Dixon (Editors)
Introduction to Building Services, second edition E.F. Curd and C.A. Howard
Introduction to Valuation, third edition D. Richmond
Marketing and Property People Owen Bevan

(continued overleaf)

Measurement of Building Services George P. Murray
Principles of Property Investment and Pricing, second edition W.D. Fraser
Project Management and Control David Day
Property Development Appraisal and Finance David Isaac
Property Finance David Isaac
Property Management – A Customer-focused Approach Gordon Edington
Property Valuation Techniques David Isaac and Terry Steley
Public Works Engineering Ivor H. Seeley
Quality Assurance in Building Alan Griffith
Quality Surveying Practice, second edition Ivor H. Seeley
Real Estate in Corporate Strategy Marion Weatherhead
Recreation Planning and Development Neil Ravenscroft
Resource Management for Construction M.R. Canter
Small Building Works Management Alan Griffith
Social Housing Management Martyn Pearl
Structural Detailing, second edition P. Newton
Sub-Contracting under the JCT Standard Forms of Building Contract Jennie Price
Urban Land Economics and Public Policy, fifth edition P.N. Balchin, G.H. Bull and J.L. Kieve
Urban Renewal – Theory and Practice Chris Couch
Value Management in Construction B. Norton and W. McElligott
1980 JCT Standard Form of Building Contract, third edition R.F. Fellows

Macmillan Building and Surveying Series
Series Standing Order ISBN 0–333–69333–7

You can receive future titles in this series as they are published by placing a standing order. Please contact your bookseller or, in the case of difficulty, write to us at the address below with your name and address, the title of the series and the ISBN quoted above.

Customer Services Department, Macmillan Distribution Ltd
Houndmills, Basingstoke, Hampshire RG21 6XS, England

Measurement of Building Services

GEORGE P. MURRAY

FRICS
Chartered Quantity Surveyor
Formerly Napier University, Edinburgh

First published 1997 by
MACMILLAN PRESS LTD
Houndmills, Basingstoke, Hampshire RG21 6XS
and London
Companies and representatives
throughout the world

ISBN 0–333–67593–2

A catalogue record for this book is available from the British Library.

This book is printed on paper suitable for recycling and made from fully managed and sustained forest sources.

10 9 8 7 6 5 4 3 2 1
06 05 04 03 02 01 00 99 98 97

Printed in Hong Kong

Contents

Preface

The principal aim of this book is to assist students studying for degree examinations in quantity surveying in the specialist area of measurement of building services within the important subject of building measurement. No previous text book has been published covering this topic under the *Standard Method of Measurement of Building Works – Seventh Edition* (SMM7) and this book aims to fill that need.

Additionally, the book should prove very helpful in the preparation of bills of quantities for building services by practising quantity surveyors or services engineers who are unfamiliar with the requirements of SMM7 and the SMM7 Measurement Code.

The book explains concisely the rules and requirements of SMM7 in relation to the measurement and billing of building services. Eleven worked examples are given in full detail in the format of draft bills of quantities to demonstrate the practical application of these rules. The worked examples are based on typical modern building services installations and are supported by detailed drawings.

To gain the most from this book, readers are recommended to have copies of SMM7 and the SMM7 Measurement Code at hand for reference while following the text and the worked examples. This advice is given as the rules from SMM7 are not generally repeated in the book, although those rules which require further explanation are dealt with in considerable depth.

After covering the general historical background of quantity surveying practice, the book leads on to a detailed explanation of the rules and techniques for measuring building services. This is followed by seven chapters covering the various specialist areas of building services from disposal systems through to alterations to existing installations. The relevant worked examples immediately follow the explanatory text within the various chapters. The final chapter covers a practical approach to producing traditional trade bills of quantities under SMM7 and details the circumstances where the adoption of such bills may be appropriate.

Edinburgh
Spring 1997

George P. Murray
FRICS

Acknowledgements

The author expresses his thanks to the Standing Joint Committee for the Standard Method of Measurement of Building Works for kind permission to quote from the *Standard Method of Measurement of Building Works: Seventh Edition (SMM7)* and the *SMM7 Measurement Code.*

Mr Edward Landor, Head of Education and Training at the Royal Institution of Chartered Surveyors is acknowledged for his kind permission to reproduce drawings from past RICS examination papers which have been adopted for some of the Worked Examples in this book.

My grateful thanks are due to Professor Ivor H. Seeley, the Series Editor, for his friendly advice and constructive criticism throughout the writing of this book. It was most reassuring to have his vast experience of technical authorship and quantity surveying practice available as required.

William Goodall and Alastair Lauener, colleagues at Napier University, are thanked for their help in the production of this book: the former for his encouragement and suggestions, and the latter for his assistance in mastering the complexities of word processing on my computer.

Malcolm Stewart of Macmillan Press Ltd is thanked for his assistance and patience in this first endeavour on the author's part.

1 General Introduction

History of Quantity Surveying

Quantity surveyors have been involved with the construction process for at least 350 years, providing the financial management of contracts and advising clients on value for money for proposed works. It is therefore rather surprising that the role of the profession has not been particularly widely recognised by the general public. This can be explained to an extent by contrasting the roles of architects and civil engineers with those of quantity surveyors. The design professions are much more publicly recognised because their creations are long-lasting, visible memorials to their endeavours, whereas the essential financial planning and management of these same projects as undertaken by quantity surveyors are not at all obvious. Over the last two decades, however, the profile of the quantity surveying profession has been raised within the public domain, partly due to the changing roles adopted by the profession, as described by Ivor H. Seeley in *Quantity Surveying Practice* (Macmillan, second edition, 1997), and the increased demand by clients for sound financial advice and management of projects. Consultant quantity surveyors are now sometimes the lead consultants for large developments, providing all professional services from inception to occupation.

In the seventeenth century quantity surveying was often a part-time occupation that provided financial reconciliation of completed building works on a measure-and-value basis by producing a final account agreeable to both client and contractor. Frequently both parties would employ independent surveyors to act for them on a fee basis to negotiate the final account. These part-time surveyors or measurers could be architects, master craftsmen or land surveyors. The Great Fire of London in 1666 caused specialisation into full-time measurement, owing to the unprecedented volume of work being undertaken, and thus provided the incentive to some quantity surveyors in London to become a profession in their own right. This process was much slower in the rest of Britain, and it was towards the middle of the nineteenth century before the profession became nationwide.

Early legal recognition of surveyors or measurers of building work was evident in Edinburgh in 1729, with the Town Council minutes documenting the Council's insistence on proof of competence before admitting 'Sworn Measurers'. These Sworn Measurers were the only surveyors whose measurements were accepted in legal proceedings in the town. This legalistic approach in Scotland to such matters also led to the first approved method of measurement of building work being published in 1773; this was known as the 'Edinburgh Mode of Measurement'. Other areas of Scotland followed with their own local modes of measurement, leading to the publication in 1918 of the first national method of measurement entitled 'The Scottish Mode for the Measurement of Building Work'.

The more *laissez-faire* approach in London and the rest of the UK may have delayed formal agreements on methods of measurement, but did lead to important innovations within the profession, such as prior costing of proposed works and preparation of bills of quantities, both demanding skilled interpretation of the drawings of the proposed works. Documentary evidence of these activities is evinced during the 1830s with the probable costings of the Houses of Parliament at Westminster and the bill of quantities for the Carlton Club in London, although earlier attempts at such professional advances almost certainly occurred. The benefits of a standard set of rules for measurement became more pressing towards the end of the nineteenth century but, owing to the disruption of the first world war, was delayed until 1922 with the publication of the first edition of 'Standard Method of Measurement of Building Works'. This document has been revised over the years, leading to the seventh edition in 1988 – SMM7 on which this book is based.

History of the Measurement of Building Services

Building services installations have a much shorter history than that of the quantity surveying profession, with the possible exceptions of rainwater disposal and land drainage which have quite early beginnings. Piped water services and sanitary drainage within buildings date from the early nineteenth century, followed by piped heating systems towards the latter half of that century, while electrical and air conditioning installations were developed in the early twentieth century. Thus quantity surveyors were in existence to measure these trades as they were introduced. However, although billing of plumbing and drainage became commonplace, the detailed measurement of mechanical and electrical services has not become so universally adopted. There are several reasons for this:

(1) Detailed design work on services installations is not sufficiently complete for billing at the tendering stage. Liaison between the client, architect and services engineer is not always adequate to

clarify the brief sufficiently early for billing, with the result that Provisional Sums are often included.

(2) Some services engineers are hostile to quantity surveyors being involved with their work. Some engineers try to insist on billing their own designs, possibly because of the aforementioned lack of completeness of the design.

(3) Very often a significant element of contractor design is incorporated into services projects, often on a performance-related specification basis. This element is difficult to incorporate in traditional bills of quantities.

(4) Quantity surveyors are not always sufficiently skilled in the technology of services installations or in the appropriate SMM7 measurement requirements. Such surveyors are usually content to agree to a lump sum approach or acquiesce to the engineer preparing the bill.

However, there has been a greater interest in including mechanical and electrical services measurement within bills in more recent years, principally because such installations have become an increasing proportion of the total cost of buildings. In modern hospitals, for instance, the various services elements can readily account for as much as 50 per cent of the total project cost. It follows that if there is no history of billing services for such projects, and as cost planning systems are usually based on detailed analysis of bills of quantities for prior projects, the cost control of 50 per cent of the work is made much more difficult. Clients are naturally concerned with the total cost of the whole project, and can reasonably expect professional cost planning and control to be exercised over all elements. Some very large cost overruns occurred in the 1970s and 1980s in projects with significant services elements, resulting in litigation and recrimination against the contractors and consultants involved. With that background it is not surprising that the detailed billing of building services is considered more favourably than in times past.

Reasons for Bills of Quantities

The very earliest bills were probably prepared in the late eighteenth century, requiring a greater degree of competence from the surveyor – that of visualising and quantifying in three dimensions from a two-dimensional plan. This skill was a considerable advance on the old measure-and-value techniques applied to already built construction. Bills of quantities became almost universal in the UK by the end of the nineteenth century. The reasons for the popularity of bills of quantities can be summarised as follows:

(1) All tendering contractors base their prices on the same information and therefore tenders are strictly comparable even if an error exists in the bill.
(2) Contractors are saved the costly exercise of each having to take off quantities for themselves. Should there be an error in their quantities, the result would be that the tender figure is too high or too low irrespective of the intended rates for items of work.
(3) Bills provide a fair basis for valuing variations and adjustments for the final account.
(4) Bills provide a convenient basis for valuation of certificated stage payments during the contract.
(5) Bills provide an approximate checklist for the contractor to order materials and other resources.
(6) Bills are essential for the preparation of detailed cost analyses for use in future cost planning.

Within the last decade, clients have been attracted to an increasing extent to alternative methods of procurement from the traditional bill of quantities approach. One method which has become popular is the 'design and build' system which transfers more risk to the contractor. This has not led to the demise of bills of quantities, because the majority of contractors still prefer to quote using a traditional bill which they commission along with the project design.

Contract Documentation

Documentation prepared for construction contracts can be separated into essential and desirable elements. Bills of quantities, while being very useful, are not contractually essential as can be gauged from the earlier sections of this chapter. The essential elements of documentation comprise three documents which define precisely what has to be provided by the contractor on the one hand and what obligations the client has on the other. These three essential documents are, in order of importance:

(1) *The conditions of contract*
These lay down the legal agreement which will be formalised between the client and the successful contractor. They cover such important matters as, for example, timing of start and finish, method and timing of payments, sequence of construction, site access, liquidate damages for delay and dispute resolution.
(2) *The contract drawings*
These provide the essential locational information, along with the

overall design and details of the building. They encompass structural, finishes and services design.

(3) *The specification*
This provides the necessary qualitative information and should explicitly define the quality of all work indicated on the drawings. Quality within building work represents a large range from cheap to expensive which the drawings alone cannot define. (It should be noted that, where a bill of quantities is produced, the specification may not be a wholly stand-alone document, as some of the specification information may be incorporated in the form of preambles to the work sections of the bill.)

In the traditional format of building contract, the three elements of documentation are normally produced on behalf of the client by consultant architects, engineers, surveyors and lawyers. In design and build contracts the contractor will usually produce the drawings and specification while the conditions of contract may be produced by either party. With design and build a fair amount of negotiation takes place on all aspects before a tailored scheme is agreed, normally with professional consultants acting for the client.

This book is not directly concerned with these three elements of contract documentation but will allude to them as necessary to explain features in the measurement examples. All aspects of contract documentation should interrelate and this is equally true when a bill of quantities is produced. The bill of quantities, by its nature, distils information from the other documents into a quantified list of all components needed to construct the project.

Bill Production Techniques

Bills of quantities comprise two main elements which complement each other, namely *items* and *quantities*. To produce accurate and useful bills, both elements must be carefully prepared in a logical sequence to avoid missing anything of importance. The writing of items requires the skills of interpreting drawings, understanding how the work will be carried out, including the operative skills involved, communication in concise technical language and knowledge of the applicable rules of measurement for the work. Calculation of the quantities for an item additionally requires the skills of mensuration, spatial concept of three dimensions from a two-dimensional drawing and a logical noting system to record the data for future use.

Two main methods of bill production have developed within the quantity surveying profession, as detailed below.

Direct Billing

This is the traditional approach, tracing its roots back to the original trade-by-trade measuring of completed work. The surveyor takes off one trade or work section at a time and works steadily through the work, producing a draft bill, usually with fully developed items and headings as well as the quantities. Once this document is calculated and checked the bill can be typed directly from this draft.

Larger projects are split between several quantity surveyors by a trade or work section basis with each preparing draft bills.

In practice, certain works must by their nature be taken off and billed direct, for example, building services and works of alteration.

The major advantage of direct billing is the simplicity of dealing only with the concept of one trade at a time. On the other hand, great care is necessary to prevent duplication or omission when several surveyors are employed in taking off one project.

Group System

The group system is a sophisticated development of the direct system, being effective for taking off quantities for whole elements of construction involving several trades. Examples of such elements are substructures, walling, finishes, roofs and components such as doors and windows, complete with the adjustments in the structure for their openings. The benefits of the system are the multiple use of quantities and the clear division of taking off tasks among a team of surveyors working on the same project. The main disadvantage of the group system is its three-stage approach, involving separately taking off, abstracting and billing, which can lead to errors in transferring data from stages.

The group system was mechanised some thirty years ago with the 'Cut and Shuffle' approach and now tends to be computerised, with several excellent software packages available.

The system does not have any relevance to the billing of building services, which by their nature are unique and not capable of being grouped. The examples in this book are therefore based on the direct billing system.

Basic Quantity Surveying Skills

Throughout this chapter, references have been made to the development of the skills of the quantity surveying profession. The basic skills of the profession are clearly required in order to measure building services effectively. It is not considered that coverage of these basic elements is part of this book and readers who require to study these skills are referred to *Building Quantities Explained* by Ivor H. Seeley, published in this series. The specific skills required for measuring building services are introduced in Chapter 2 of this book.

Metric Units

The British construction industry converted to the metric system in the late 1960s, adopting the SI system of units. This system uses only millimetres and metres for dimensions, which has the advantage, because of the factor of 1000 between them, of limiting errors caused by mistaking units. For example, a manhole cover 750 × 600 must be in millimetres as that size in metres would represent an impossibly large manhole cover. Equally, a manhole cover quoted as 0.750 × 0.600 must be in metres, as in millimetres it would be impossibly small. Thus the need to quote 'm' for metres and 'mm' for millimetres is obviated in most cases, and on the introduction of the SI system the British Standards Institution strongly advocated the omission of metric symbols on drawings, specifications and bills of quantities.

However, the quantity surveying profession has not universally adopted the above advice and many bills continue to be produced with the symbols included. In fact, the specimen bill pages contained in Appendix 4 of the SMM7 Measurement Code quote 'mm' as appropriate throughout the four sample pages.

The Standard Method of Measurement of Building Works: Seventh Edition (SMM7) uses the metric symbols throughout for categories required by the rules of measurement, for example:

Rule P31.20.1 Cutting or forming holes for other services installations, ducts, categories:

(1) girth not exceeding 1.00 m;
(2) girth over 1.00 but not exceeding 2.00 m;
(3) and thereafter in 1.00 m stages.

On balance, it was considered that in this book it would be prudent to include the metric symbols where appropriate and allow readers to judge whether to include them or omit them. A sample item follows in both formats to illustrate the point:

Item	450 mm Diameter inspection chamber comprising chamber base with blanking plugs set on grade 20 *in situ* concrete foundation 100 mm thick, cast iron cover and frame and chamber risers to suit 820 mm invert	nr	1

Item	450 Diameter inspection chamber comprising chamber base with blanking plugs set on grade 20 *in situ* concrete foundation 100 thick, cast iron cover and frame and chamber risers to suit 820 invert	nr	1

2 Rules and Techniques for Measurement of Services

Introduction

This book is based on the rules of measurement under the Standard Method of Measurement of Building Works: Seventh Edition (SMM7) and the SMM7 Measurement Code. SMM7 has been accepted throughout the United Kingdom, but has not been generally adopted in the Republic of Ireland or elsewhere in the world. The principles of good measurement practice contained in SMM7 and demonstrated by the examples contained in this book may, however, be applied to any set of standard measurement conventions in use elsewhere in the world.

The adoption of a standard set of measurement rules has great advantages for the construction industry generally. More specifically, advantages are gained for tendering contractors and for quantity surveyors acting both for clients and contractors. These advantages are summarised below with reference to SMM7:

(1) Bill items properly framed under the SMM have standard inclusions automatically deemed to be included which tendering contractors should take account of in estimating their rates. A good example is SMM7 General Rule 4.6 which deems 'labour, materials, assembling, plant, waste, square cutting, establishment and overhead charges and profit' to be included in all items unless otherwise specifically stated in a bill of quantities.

This helpful rule completely removes the necessity for repetitive use of phrases such as 'supply, fit and fix' which would otherwise be required and are not nearly so inclusive.

(2) Following from (1) above, *extra* elements of cost not included in the rules should normally generate additional items to be priced by the contractor.

For example in SMM7 – Section R Disposal Systems – Section R10/R11 Rule C3 deems joints on pipes in running lengths to be included (having been specified in the item description), but Rule D2 requires special joints and connections which differ from those generally occurring in the running length to be enumerated as extra over the pipe in which they occur.

(3) Following from (1) and (2) above, contractors should have confidence in pricing items within bills of quantities which are framed under the clear guidance of SMM7.
(4) Adjustment of variations in the final account of a project can be based on the knowledge of what ought to have been included in the original bill item rate.

These valuable features of SMM7 serve to improve the usefulness of bills of quantities and their acceptance within the construction industry. While all the foregoing benefits exist, they can be negated or diminished by differing interpretations of the rules and often in the specific case of building services by a lack of appreciation of the technology of the work being measured. This book attempts to explain the former and encourage greater awareness of the latter.

SMM7 General Rules

The use of SMM7 has been justified above but some general principles which are common to all work should be highlighted before proceeding to the specifics of building services. The most important rule in the whole SMM is General Rule 1, which should be kept in mind whenever billing is being done. This rule sets the ethos and standards for quantity surveyors to aim for in measurement practice, namely:

(1) Produce bills of quantities which fully describe and accurately represent the quantity and quality of the work to be carried out.
(2) Give more detailed information than is required by the rules where necessary in order to define the precise nature and extent of the required work.

In order to comply fully with these worthy objectives, surveyors must be professionally competent in the technology and measurement practices of the work in hand.

Using SMM7 successfully also requires a good working knowledge of all the other General Rules.

SMM7 Measurement Code and SMM7

The SMM7 Measurement Code is mentioned at the beginning of this chapter and deserves clarification as to its status and validity in relation to SMM7. In the Preface to the Code it is stated that it is non-mandatory and intended to be read in conjunction with SMM7. The Code's principal objectives are to encourage good practice in the measurement of building works and preparation of bills of quantities, and to encourage uniform interpretation and use of SMM7. These worthy objectives are very much in line with those of this book, and references to the Code are included in the Worked Examples wherever appropriate.

Before starting to measure work within any particular SMM7 Work Section, it is always advisable to be familiar with any recommendations contained in the Code for that topic. There are a considerable number of references to building services within the Code.

Building Services in SMM7

Building services are comprehensively covered by the rules contained in SMM7, but are not always readily accessed via the index. *Plumbing work,* for example, is not featured in the index because the contents are arranged under the recommendations of the Committee for, 'Coordinated Project Information' (CPI) and the 'Common Arrangement of Work Sections' (CAWS). Both CPI and CAWS were developed very much with the documentation of large complex projects in mind and they are not particularly helpful in assisting the clarity of documentation intended for small to medium sized works. Plumbing work has not of course been abolished by CPI and CAWS, but exists under various headings such as 'Disposal Systems' and 'Piped Supply Systems'.

SMM7 has two distinct ways of handling the rules for Building Services. The first system is straightforward and comprises direct rules within their own sections, which is in common with the great majority of other sections within SMM7. These direct sections cover four work sections in two sets of rules:

R10 Rainwater pipework/gutters; R11 Foul drainage above ground

R12 Drainage below ground; R13 Land drainage

The second system is a two-stage approach with work sections contained in Appendix B of SMM7 'Classification of mechanical and electrical services' and the rules contained in Section Y 'Mechanical and electrical services measurement':

Y10–59 cover the various types of mechanical services measurement

Y60–92 cover the various types of electrical services measurement

The reason for the two-stage approach is the very large number of work sections all having a common core of measurement rules, which under the direct system would have resulted in several pages of headings before the rules commenced. The Appendix B strategy gives a much clearer and neater layout to the SMM. It could be argued that Sections R10–13 should have been incorporated into the Appendix B/ Section Y system for consistency, but this has not been done.

When preparing bills of quantities for services included in Appendix B it is essential to incorporate appropriate headings from both Appendix B and Section Y, so that the tendering contractor is not left in any doubt as to the nature of the work and the actual rules applying to specific items. Each heading within Appendix B will generate a subsection of the bill. For instance, examples of such subsections of a typical electrical installations bill would be: V20 LV Distribution, V21 General Lighting, V22 General LV Power, V40 Emergency Lighting. Within each subsection the items would be further broken down into the appropriate Section Y topics as in the following example:

V20 LV Distribution	(Appendix B) – Subsection heading
Y60 Cable trunking	(Section Y) – Heading to identify specific rules
(Bill items detailing trunking requirements)	
Y61 LV Cables and wiring	(Section Y) – Heading to identify specific rules
(Bill items detailing cables and wiring)	
Y80 Earthing	(Section Y) – Heading to identify specific rules
(Bill items detailing earthing requirements)	

Any other approach to billing is likely to cause confusion and could lead to future argument.

Basic Principles of Services Measurement

Measurement of building services has frequently been avoided by quantity surveyors over the years, principally because of a lack of understanding of the technology involved and difficulty in interpreting the drawings. This was particularly true in the case of heating, ventilating and electrical works. Building services may however be simplified into three functional elements applying to all types of supply installation, namely:

Source: Examples – Boilers, Air handling equipment, Electrical main switchgear

Distribution: Examples – Piping, Ducting, Wiring, Conduit

Outlets: Examples – Radiators, Air grilles, Lighting points, Power outlets.

Source and **Outlet** elements are usually enumerated, whereas **Distribution** elements are generally measured by length. If installations are considered in this way then the approach to measurement is clearer and the chance of missing essential elements of the work is reduced. The concept can be illustrated diagrammatically:

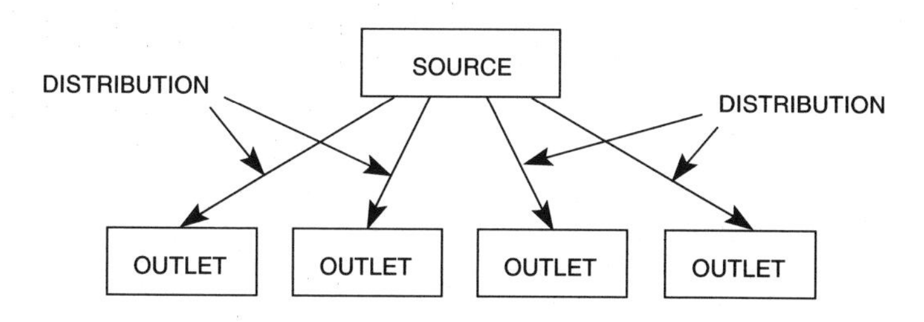

The above concept generally applies to all supply services installations, but a very similar concept can be applied to disposal services installations with the following modified diagram:

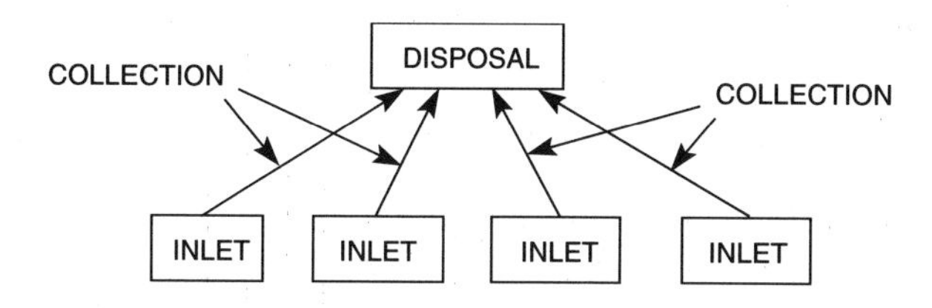

Disposal: Examples – Main drain connection, Sewage pump, River outfall, Soakaway

Collection: Examples – Piping, Land drain, Guttering

Inlets: Examples – Sanitary appliances, Rainwater heads, Gullies.

Disposal and **Inlet** elements are usually enumerated, whereas **Collection** elements are generally measured by length.

Techniques for Taking Off Building Services

For most of the work comprising building services the techniques required for taking off quantities are the straightforward approach applying to any other work, but there are certain elements where different techniques are appropriate and further explanation is warranted.

Pipelines
Pipelines occur in many building services, from simple plumbing to complex specialist installations. Except for the most basic installations, pipes occur with a variety of sizes, materials, supports, fittings and accessories in various locations. The pipes for the plumbing installation of a large building that includes these features can easily generate over 100 items. Attempting to take off quantities for pipes from the drawings directly inclusive of all features is virtually impossible. The logical approach to the problem is to take off the quantities for a section or branch of pipeline complete with all sizes of pipe and features – commencing at one end of the pipe and methodically working to the other end by noting the measurements on an abstract sheet as a first stage. All sections and branches of pipes are thus abstracted with quantities totalled and checked, and then billed from the abstracts as a second stage. A typical sample of an abstract sheet is included at the end of this chapter.

Repetitive Components
Some components of services installations are repeated throughout the project but vary in one or two features only. Examples would include manholes in drainage works, where the depths and connections could vary between the units but all would be constructed to a standard design. Heating radiators are another example, where they might only vary by size, output and fixing but otherwise be the same. Such items are more readily billed from a schedule of components which is a tabular presentation of the necessary data. Engineers quite often produce these schedules, but failing that it may be more efficient for the surveyor to produce them as an aid to billing. A typical sample of a schedule is included at the end of this chapter.

A specialist abstract is usually produced by the engineer for electrical installations and this tabular data is known as a 'Distribution Sheet' which is very useful when billing the work. Examples of distribution sheets are contained in Chapter 7.

Builder's Work in New Construction

Builder's work in connection with building services is principally covered by SMM7 – Sections P30/31, respectively 'Trenches/Pipeways/Pits for buried engineering services' and 'Holes/Chases/Covers/Supports for services'. Builder's work is to be identified under separate headings for plumbing, mechanical and electrical installations – Rule P30/31 M2. Surveyors are still left with the options of billing builder's work within a separate bill, within the Builder trade bill, or within the bills for individual services, provided that appropriate headings are employed. The choice of where to bill these items should be made according to the size of the project and the contractual arrangements proposed. In this book, for ease of location, builder's work will be covered within the bills for the individual services.

Other builder's work not identified within Sections P30/31 is to be given in accordance with the rules of the appropriate work section – Rule P30/31 M1. In such cases it is best to bill this work within the relevant work section. An example of such work would be holing proprietary roof decking for pipes, where this would be most appropriately measured in the bill along with the decking – SMM7 – Section J43 Rule 21.

As there is some difficulty in deciding which set of rules applies, further explanation for the commonly occurring features is required, as discussed below.

Trenches

Trenches for all types of buried services are measured under P30.1–7, with the exception of underground drain trenches which have their own rules under R12/13.1–7. The reason for this approach is not clear, as both sets of rules are virtually the same throughout.

Holes for Cables, Ducts and Pipes

Holes, mortises, sinkings and chases for electrical installations are measured in a straightforward manner, being enumerated on a 'points' basis under P31.19. These items cover all operations required in the formation of ways for electrical work in *new build* projects.

Rule P31.20 covers holes for services installations other than electrical work, and includes holes for ducts and pipes. These are each enumerated in three categories of size and can include making good. This appears simple as it stands, but there are many other instances of rules for pipe holes in many other locations throughout SMM7. The rules covering holing work of other trades for services (other than electrical services) are not completely clear, but can be simplified as three possibilities:

(1) Holes formed after the building structure is built but before finishes (if any) are applied – measured as P31.20.
Examples
Holes through brick/block/stone walling.
(Beware of possible confusion caused by Rule C1(c) in both F10 – Brick/block walling and F20/21/22 – Stone walling, which states that 'holes are deemed included' in walling. It is not considered that this refers to holes for services but only to holes formed during construction for air vents and the like.)
(2) Holes formed in building structure during construction – measured in relevant work section.
Examples
Holes formed in *in situ* concrete by formwork – E20.27.
Holes cut in *in situ* concrete by specialist – E41.11.
(3) Holes formed in structure or finishings around pipes or ducts where the 'roughing' work of the service installation has generally already been installed – measured in relevant work section of the structure or finish.
Examples
Holes in timber flooring/panelling – K13/20/21.10.
Holes in sheet tanking or built-up roofing – J40/41.21.
Holes in slate or tile roofing – H60–64.11.

Builder's Work in Existing Buildings

This topic is fully covered in Chapter 9, Measurement of Services in Existing Buildings.

SAMPLE SCHEDULE
Drainage Scheme – Golf Clubhouse Project

MANHOLE SCHEDULE

MH Nr	Internal size	Wall thickness	Invert level	Original ground level	Cover level	Exc'n depth	MH depth to invert	Features	Notes
1 (Drop)	1200 × 1200	$1\frac{1}{2}$ brk	9.880	13.350	13.560	3.670	3.680	Upper MH size 750 × 700	200 *in situ* conc base MH steps @ 300 ctrs
2	750 × 1200	$1\frac{1}{2}$ brk	11.160	14.260	14.260	3.300	3.100	Upper MH size 610 × 610	Ditto
3	750 × 750	1 brk	11.340	12.550	12.900	1.410	1.560	Nil	Ditto

SAMPLE ABSTRACT SHEETS
Heating Installation – Golf Clubhouse Project
Abstract Nr 1
25 mm MS screwed pipework BS 1387 medium weight

Location	Timber backgrd pipe rings	Brick backgrd pipe rings	Fixings separate	Laid in floor duct pipe clip	Elbows	Demount couplings	Formed bends	Formed offsets	Pipe sleeves half bk walls
Kitchen	1.85	1.65			2	1			1
Servery		8.00		1.80	5	2		1	
Lounge	2.60		1.90	2.55	3	3	2	2	1
Corridor		3.70	2.00		2	1	1	1	1
Totals	4.45	13.35	3.90	4.35	12	7	3	4	3

Abstract Nr 2
32 mm MS screwed pipework BS 1387 medium weight

Location	Timber backgrd pipe rings	Brick backgrd pipe rings	Fixings separate	Elbows	Demount couplings	Demount elbows	Formed bends	Pipe sleeves half bk walls	Pipe sleeves one bk walls
Toilets	2/5.85			2	2	2	2	1	1
Kitchen	2/1.15	2.60	0.90	1	1	2	1		1
Servery	1.85		1.00	2		2		1	
Locker Room	2/4.40		2/2.35	4	2	2	4	2	2
Lounge		6.50	3.75	2	3	2			2
Totals	24.65	9.10	10.35	11	8	10	7	4	6

3 Measurement of Disposal Systems

Introduction

Disposal systems comprise such installations as rainwater disposal, soil and waste disposal, underground drainage (surface water or foul) and land drainage. In Chapter 2 the general approach to billing and measurement of building services was covered, indicating the three basic components of disposal services installations – namely: disposal, collection and inlets. Examples of the respective components are:

Disposal	**Collection**	**Inlets**
Connection to sewer	Pipework	Sanitary appliances
Manholes	Land drains	Rainwater heads
Septic tank	Guttering	Gullies

Measurement Rules

Disposal systems are largely covered by SMM7 Rules R10–13, which are arranged with a common set of rules for above ground installations and another common set of rules for below ground installations, as follows:

Above ground	R10 – Rainwater pipework/gutters
	R11 – Foul drainage above ground
Below ground	R12 – Drainage below ground
	R13 – Land drainage

These sections follow the normal direct pattern of rules of SMM7 and, as discussed in Chapter 2, these works are not involved in the Appendix B/Section Y approach. The only exceptions are the very specialised disposal systems covered in Appendix B: R14–33 involving such topics as laboratory drainage, sewage pumping and incineration plant.

Sanitary appliances/fittings are dealt with separately from the disposal system rules in SMM7, and are listed in Appendix A of the Standard Method. This merits further explanation later in this chapter.

'**Disposal**' items may be simply enumerated as, for example, SMM7 Rule R12.16 – Connecting to Local Authority's sewer, but more often the requirement is for a detailed measurement in accordance with SMM7 Rules R12/13.11–15 covering manholes, inspection chambers, soakaways, cesspits and septic tanks. Measurement rule M8 states that excavation, concrete, formwork, brickwork, rendered coatings and other work are to be measured in accordance with the rules for the appropriate Work Sections. Although this level of detail can be justified for the larger chambers, such as septic tanks, this requirement of the SMM seems over-detailed for the smaller, less expensive and simpler features. Manholes, for example, if built in traditional brick and concrete construction, can generate around twenty items under SMM7, which contrasts with single-item enumeration per manhole in civil engineering practice. Quantity surveyors may well opt for a more straightforward approach in professional practice by specific exception from the SMM, but the examples in this book will be measured strictly in accordance with the rules of SMM7.

Preformed chambers are, however, enumerated under SMM7 Rules R12/13.11–15.14, which cover the factory-produced, generally moulded plastic inspection chambers and septic tanks. The SMM is not clear whether excavations and any incidental concrete bedding should be included in such an enumerated item, but it would seem sensible to adopt that approach.

'**Collection**' items are generally measured by length; for example, SMM7 Rules R10/11.1 pipes (above ground), R12/13.8 pipes (below ground) and R10.10 gutters. Labours and fittings on pipes and gutters are enumerated as 'extra over the work in which they occur' as R10/11.2, R12/13.9 and R10.11. Ancillaries or accessories to pipes, on the other hand, are enumerated as *full value* as R10/11.6 and R12/13.10, and thus it is very necessary to be clear on the distinction between fittings and ancillaries/accessories. Examples of pipe and gutter fittings are given in the Measurement Code under Section R, while pipe ancillaries are listed in SMM7 Rule R10/11.6 (some of the pipe ancillaries and accessories are listed as 'Inlets' in the general approach to billing adopted in his book, but this does not effect the resultant bill in any way), and drain pipe accessories are defined in Definition Rule D6 of SMM7 Rules R12/13.

'**Inlets**' such as gullies and rainwater heads are enumerated, being classed as pipe ancillaries or accessories in SMM7, and are measured as explained in the previous paragraph. The other common type of 'inlets' are Sanitary appliances/fittings, which are also enumerated but are dealt with in a special way by SMM7 and thus deserve a separate explanation.

Sanitary Appliances/Fittings

SMM7 deals with sanitary ware in a similar manner to the majority of building services installations (other than Rules R10–13), by having an appendix to cover the multiplicity of types and a common set of measurement rules. In this case, SMM7 Appendix A – Section N13 lists all the likely types of sanitary appliances/fittings while the measurement rules are contained in Section N Furniture/Equipment – N13. The specific rule is SMM7 – N13.4 'Fittings, equipment and appliances associated with services', which enumerates such items requiring relevant specification information and fixing details.

Coverage Rule C1 states 'Providing everything necessary for jointing is deemed to be included', which is a provision open to misinterpretation. The most likely intention of SMM7 is that all jointing *within* the sanitary fitting is deemed to be included, for example:

WC cisterns deemed to be jointed to WC pans
Waste outlets deemed to be jointed into sinks, basins or baths.

Jointing requirements to pipes *outside* the sanitary fittings are impossible to predict from the bill item and thus unfair to be deemed to be included. It is very possible that identical sanitary appliances in a project could be jointed to soil, waste, supply or overflow pipes of differing materials and diameters by differing methods of jointing. For that reason it seems unlikely that SMM7 intended contractors merely to guess at the cost implications of these variables. Nevertheless, the rule could cause confusion and for this reason an explanatory preamble is a worthwhile inclusion in bills containing sanitary appliances. Suitable wording for this clarification would be:

'All joints and connections between supply, overflow, waste or soil piping and the under noted sanitary appliances/fittings have been measured with the appropriate pipes'.

This preamble is included in the worked examples that follow.

WORKED EXAMPLE

The following **Worked Example 3/1** will demonstrate the application of the various Rules from SMM7 to typical disposal installations. Readers should refer to SMM7 and the Measurement Code while following this example, which will be cross-referenced where appropriate.

Worked example 3/1 is based on a detached two-storey house incorporating disposal systems for rainwater, soil and waste, below ground drainage and land drainage.

This worked example is set out in draft bill format, which clarifies not only the SMM7 measurement requirements but also the various ancillary matters such as testing and sundries, which may not appear clearly in a basic take-off.

Worked Example 3/1: Detached House – Disposal Systems

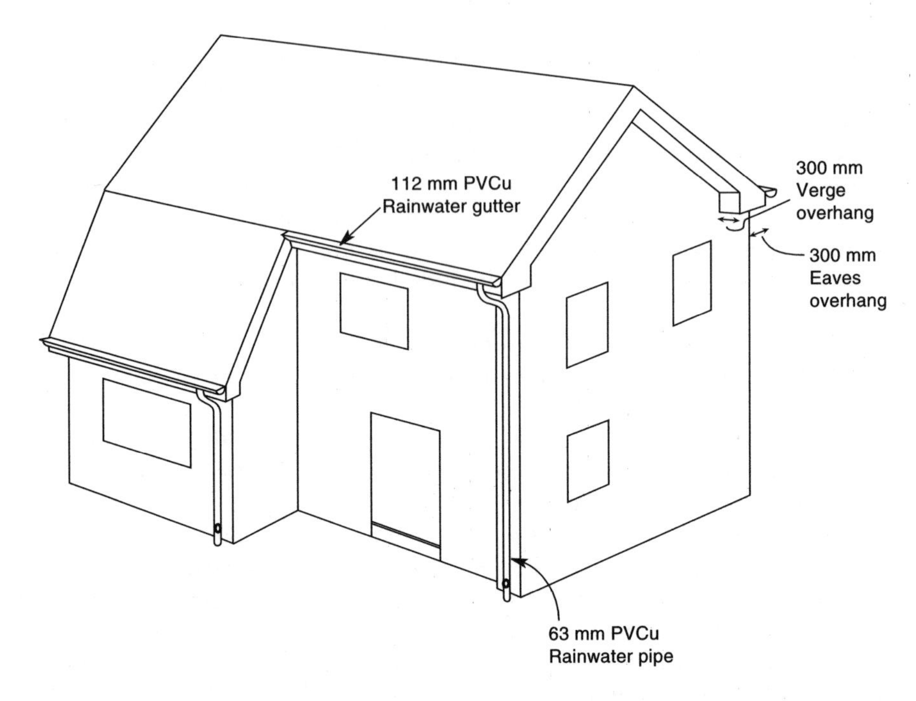

ISOMETRIC SKETCH SHOWING RAINWATER INSTALLATIONS

Brief Specification

Rainwater Goods: to be grey PVCu fixed with brass screws.

Foul Drainage Above Ground: generally grey PVCu soil piping for single stack plumbing except where below GF level to be cast iron to BS 460. Waste piping to be polypropylene with 'O' ring joints.

Drainage Below Ground: surface water drains to be PVCu piping with polypropylene couplings and PVCu inspection chambers. Foul drains to be vitrified clay piping with polypropylene couplings and brick built manholes. Land drains to be PVCu perforated flexible piping.

3/1/1

Worked Example 3/1: Detached House – Disposal Systems

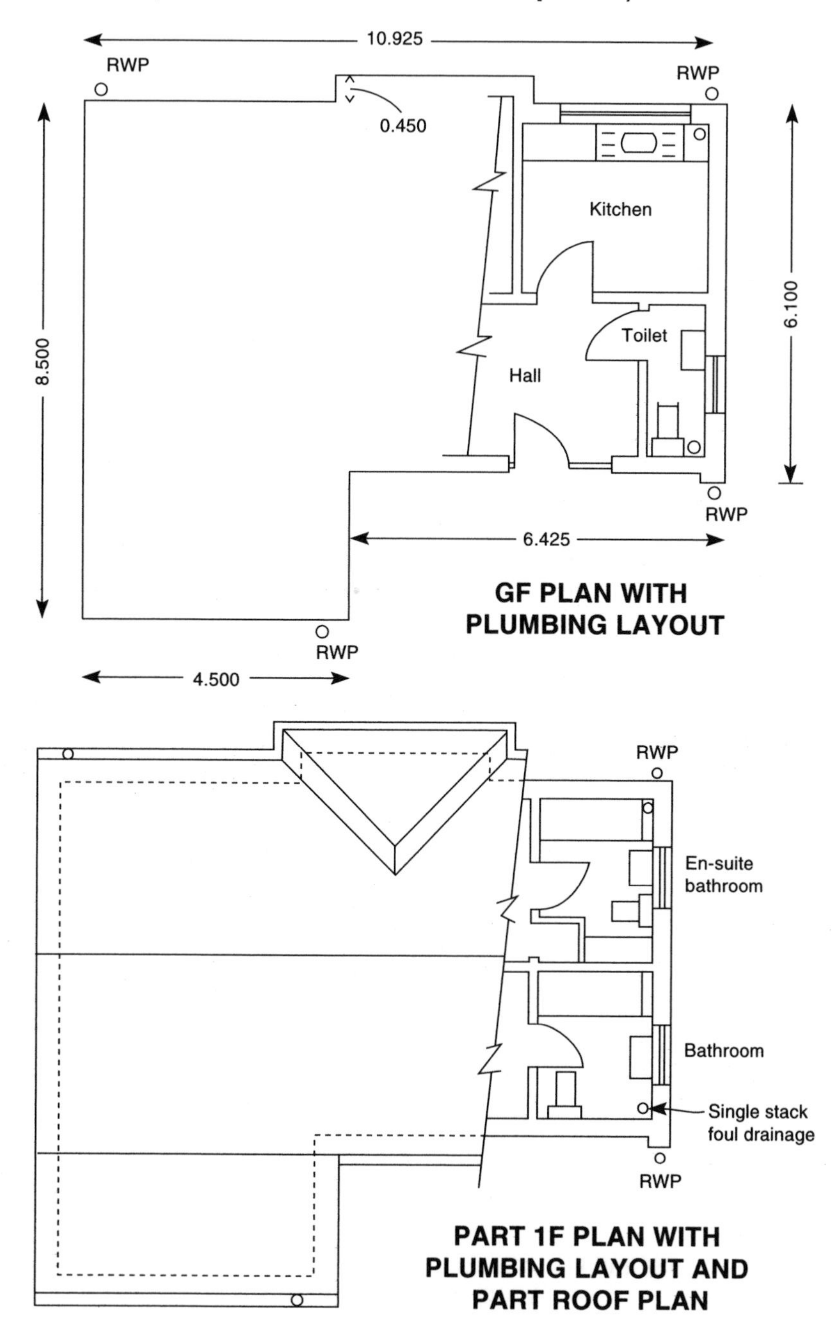

3/1/2

Worked Example 3/1: Detached House – Disposal Systems

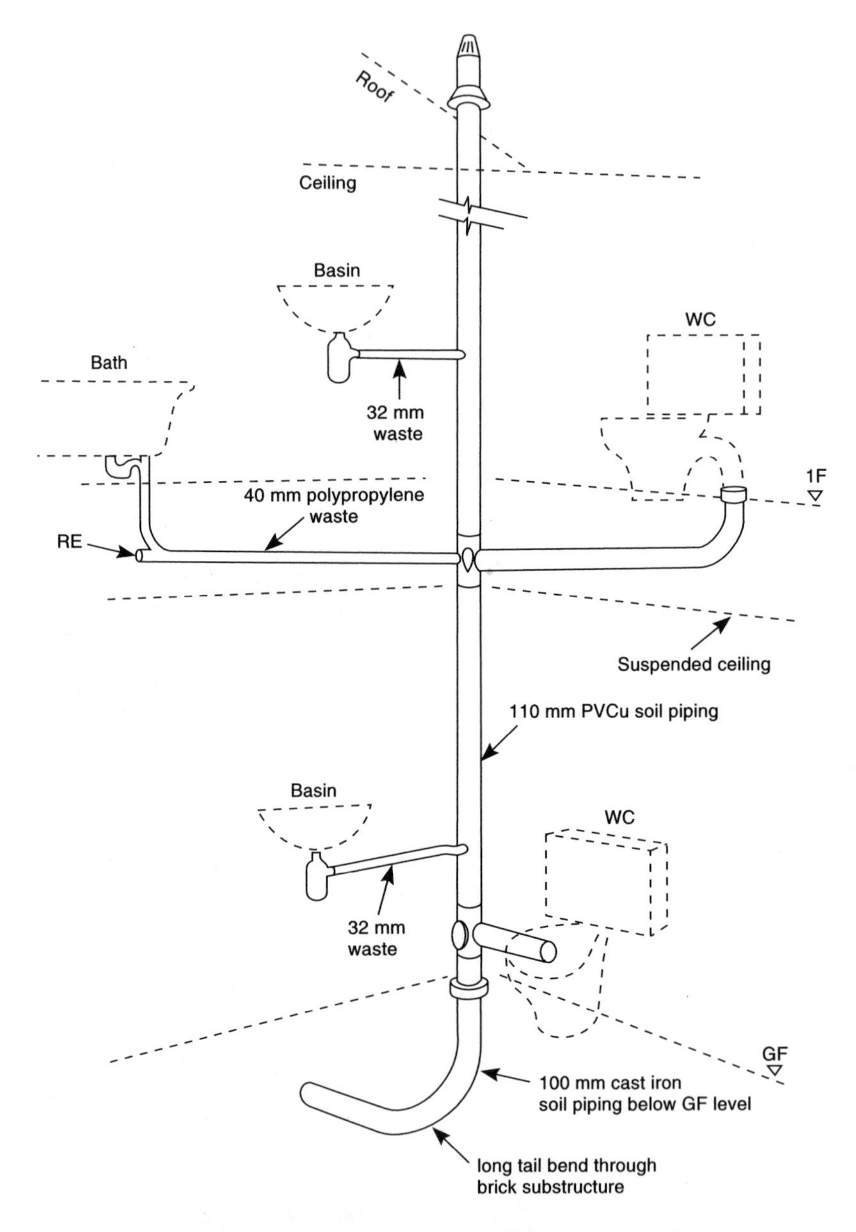

ISOMETRIC SKETCH OF SINGLE STACK IN SOUTH-EAST CORNER

3/1/3

Worked Example 3/1: Detached House – Disposal Systems

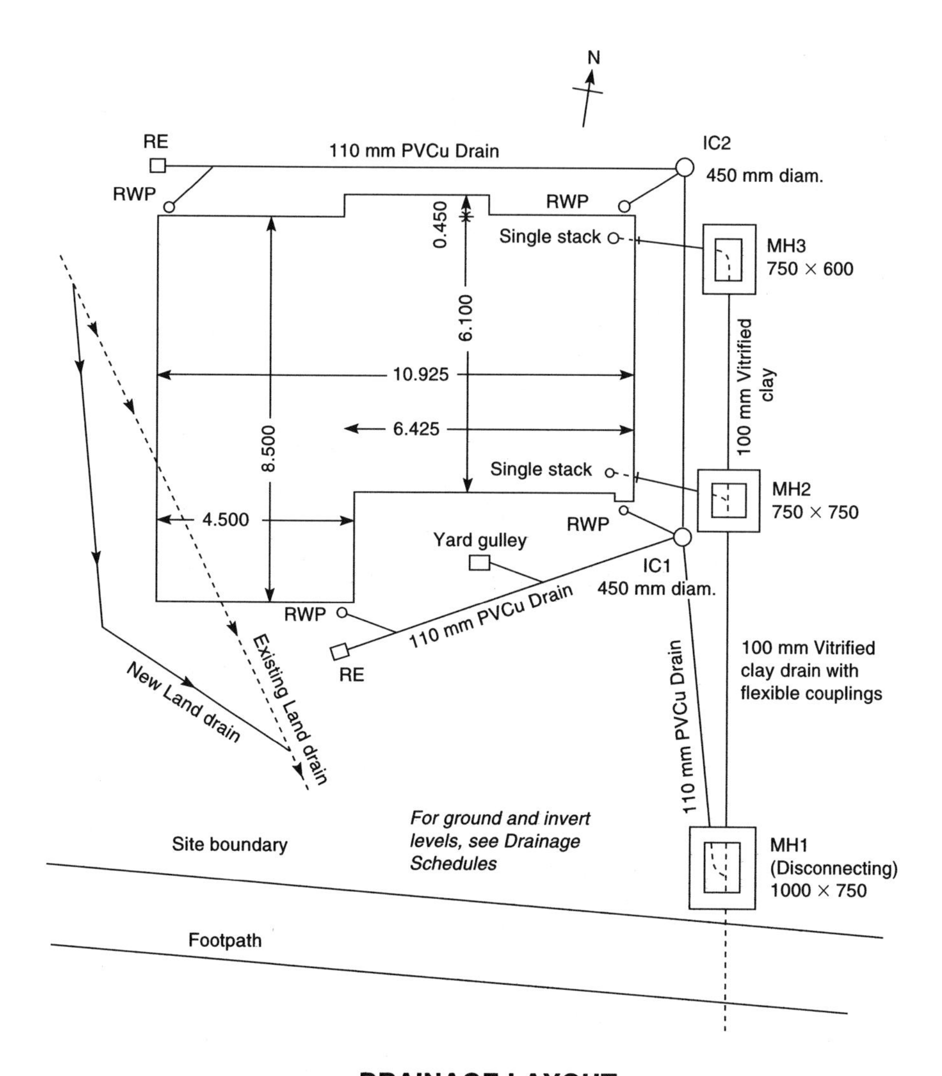

DRAINAGE LAYOUT

3/1/4

Worked Example 3/1: Detached House – Disposal Systems

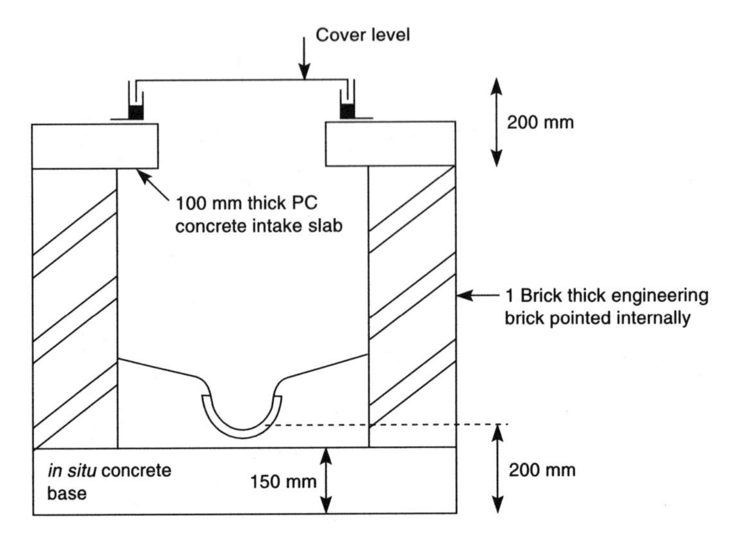

TYPICAL MANHOLE DETAILS

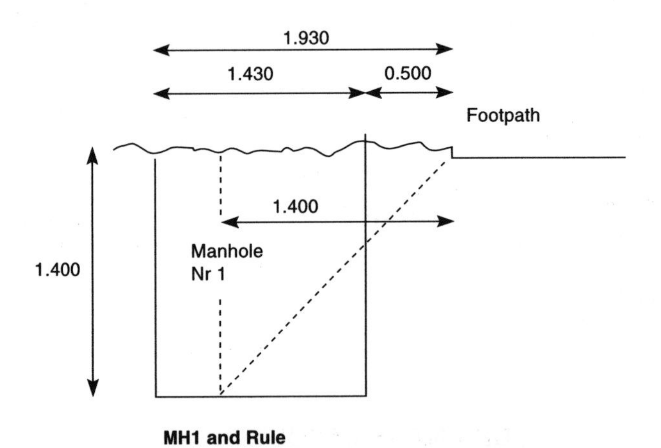

MH1 and Rule
D20.7.*.*.4
Earthwork support next to roadway – refer to Measurement Code p20

DRAINAGE DETAILS

3/1/5

SURFACE WATER DRAINAGE SCHEDULE						
Location	Ground Level	Cover Level	Invert Level	75 mm Below Invert	Trench Depth	Depth of IC
MH1 Disconnecting	21.650	21.800	20.700	20.625	1.025	N/A
IC1	21.700	21.800	20.890	20.815	0.885	0.910
IC2	21.800	21.900	21.080	21.005	0.795	0.820
RE Front	21.800	21.890	21.060	20.985	0.815	N/A
RE Rear	21.850	21.900	21.330	21.255	0.595	N/A

SURFACE WATER DRAINAGE TRENCH DEPTHS				
Trench Run	Trench Depth Range		Average Depth	SMM7 Category
MH1–IC1	1.025	0.885	0.955	N/E 1.00
IC1–IC2	0.885	0.795	0.840	N/E 1.00
IC1–RE Front	0.885	0.815	0.850	N/E 1.00
IC2–RE Rear	0.795	0.595	0.695	N/E 0.75
Branch IC1–RWP	0.885	0.500	0.693	N/E 0.75
Branch RE–RWP Front	0.815	0.500	0.658	N/E 0.75
Branch Yard Gully	0.850*	0.650	0.750	N/E 0.75
Branch IC2–RWP	0.795	0.500	0.648	N/E 0.75
Branch RE–RWP Rear	0.595	0.500	0.548	N/E 0.75

* Interpolated

FOUL DRAINAGE MANHOLE SCHEDULE									
Manhole Nr	Internal Size	Wall Thick-ness	Overall Size	Ground Level	Cover Level	Invert Level	Adjacent Trench Depth	Depth to MH Base	Total Built Height of MH
MH1 Disconnecting	1000 × 750	1B	1430 × 1180	21.650	21.800	20.450	1.300	1.400	1.550
MH2	750 × 750	1B	1180 × 1180	21.700	21.800	20.650	1.150	1.250	1.350
MH3	750 × 600	1B	1180 × 1030	21.800	21.900	20.790	1.110	1.210	1.310

FOUL DRAINAGE TRENCH DEPTHS				
Trench Run	Trench Depth Range		Average Depth	SMM7 Category
MH1–MH2	1.300	1.150	1.230	N/E 1.25
MH2–MH3	1.150	1.110	1.130	N/E 1.25
Branch at MH2	1.150	0.850	1.000	N/E 1.00
Branch at MH3	1.110	0.850	0.975	N/E 1.00

WORKED EXAMPLE 3/1				SHEET NR 1
Commentary	Item Nr	Description	Unit	Qty
		DETACHED HOUSE		
Sequential order of bills		*BILL NR 8*		
Common Arrangement of Work Section heading		*R – DISPOSAL SYSTEMS*		
		Information provided		
This note covers the requirements of Rule P1 of sections R10–13 and N13. Placing this first removes the need for repetition within the various subsections of the bill.		The following represents the installation of rainwater, foul drainage above ground, drainage below ground and sanitary appliances as detailed on Plans 3/1/1–5.		
		R10 Rainwater Pipework/Gutters		
R10.1.*.1 R10/C3		*Grey PVCu rainwater pipework with socketed joints*		
R10.1.1.1.1 'straight' will apply to 99% of pipes and SMM7 would have benefited from a 'deemed to be straight' clause.	1/1	Pipes, straight, 63 mm diameter, with plastic pipe clips at not exceeding 2.00 m centres, brass screwed and plugged to masonry	m	21
Pipe lengths measured over all fittings Rule M1.		2 storey ht 3/5.85 1 storey ht 3.15		
R10.2.3.2 Pipe fittings not exceeding 65 mm diam. simplified measurement.		Extra over 63 mm pipe:		
	1/2	Fittings with two ends	nr	4
		Eaves offsets 4		
R10.2.3.2.1	1/3	Fittings with two ends and screwed inspection door	nr	4
		Each pipe 4		
R10.2.2.1.1 R10/D2	1/4	Special connections to 110 mm diam. plastic underground drain with non-concentric plastic connector	nr	4
		Each pipe 4		
		Grey PVCu rainwater gutter with plain ends and union brackets		
R10.10.1.1.1 R10/M6 & C9	1/5	Gutters, straight, half round 112 mm, with plastic fascia brackets at not exceeding 0.90 m centres brass screwed to timber	m	24
Allowing for 300 mm eaves and verge projections to brick sizes and measured over all fittings.		Front 1 storey 5.10 Front 2 storey 6.43 Rear 11.53 Rear projections 2/0.45		

WORKED EXAMPLE 3/1			SHEET NR 2	
Commentary	Item Nr	Description	Unit	Qty
R10.11.2.1		Extra over 112 mm gutter:		
	2/1	Stop ends	nr	6
		Front 4 Rear 2		
	2/2	Running outlets	nr	4
		Front 2 Rear 2		
	2/3	Right angles	nr	4
		Rear projection 4		
		R11 Foul Drainage Above Ground		
This measurement represents the single stack system in the south-east corner only as a sample of this work.				
R11.1.1.1 plus locational information in line with requirements of R11.1.*.*.2&3		*Cast iron soil pipes to BS 460, heavy quality coated, with socketed joints, jointed with rope yarn and molten lead, work below ground floor level in substructure*		
'straight' will apply to 99% of pipes and SMM7 would have benefited from a 'deemed to be straight' clause.	2/4	Pipes, straight, 100 mm diameter with MI hinged holderbat fixings at not exceeding 2.00 m centres cut and pinned into masonry	m	1
		girth incl. bend 1.35		
R11.2.*.*		Extra over 100 mm pipe:		
Pipe fittings over 65 mm diam. given in detail R11.2.4.5	2/5	Long tail large radius bend, 500 mm long	nr	1
Most of length of item 2/4 is bend		Through wall 1		
R11.2.2.1.1 R11/D2	2/6	Special connection to 100 mm diam. clay underground drain with plastic sleeve connector	nr	1
		1		
R11.1.*.1		*Grey PVCu soil pipes with socketed joints, jointed with pushfit seal*		
R11.1.1.1.1	2/7	Pipes, straight, 110 mm diameter, with GMS bolted pipe clips at not exceeding 2.00 m centres rustproof screwed and plugged to masonry	m	7
Ground floor to u/s of vent grating		Full ht 6.23 GF branch only 0.50		

WORKED EXAMPLE 3/1		SHEET NR 3		
Commentary	Item Nr	Description	Unit	Qty
R11.7 requires pipe supports which differ from those given and shall be enumerated separately – it follows that the pipe item will not include fixings.	3/1	Pipes, straight, 110 mm diameter, fixings measured separately	m	2
		1F branch 1.50		
R11.2.*.*		Extra over 110 mm pipe:		
R11.2.4.5	3/2	Bend	nr	1
		1F branch 1		
R11.2.4.5.1	3/3	Branches with screwed inspection door	nr	2
		GF 1 1F 1		
R11.2.2.1.1	3/4	Special connection to 100 mm diam. CI soil pipe with plastic connector set in CI socket with mastic	nr	1
		GF 1		
R11.2.2.1.1	3/5	Special connection to spigot of WC with flexible plastic connector	nr	1
		1F 1		
	3/6	Special connection as last item but bent	nr	1
		GF 1		
R11.5.1.1 R11/C7 Boss connections are full value and are deemed to include perforating the pipe.	3/7	Boss connection utilising cast on boss of previously measured fitting complete with solvent welded adaptor for 'O' ring 40 mm polypropylene pipe	nr	1
		1F bath 1		
	3/8	Boss connection solvent welded to pipe with 'O' ring connection for 32 mm polypropylene pipe	nr	2
		GF basin 1 1F basin 1		
R11.6.4.1 Ancillaries are full value and therefore excluded from pipe length measured in item 2/7. R11/C8 Cutting and jointing to pipe deemed included.	3/9	Pipework ancillaries, 110 mm terminal vent cowl	nr	1
		Roof 1		
R11.6.5.1.1	3/10	Pipework ancillaries, 110 mm plastic roof flashing	nr	1
		Roof 1		

WORKED EXAMPLE 3/1			SHEET NR 4	
Commentary	Item Nr	Description	Unit	Qty
R11.7 Special pipe support	4/1	Fabricated pipe support for 110 mm pipe comprising GMS bolted pipe clip galvanised bolted to GMS hanger 20 × 3 × 300 mm long, twice rustproof screwed to timber	nr	1
		1F branch 1		
R11.1.1.1		*Polypropylene waste pipes with push fit 'O' ring straight couplings*		
R11.1.1.1.1	4/2	Pipes, straight, 32 mm diameter, with plastic pipe clips at 0.60 m centres brass screwed and plugged to masonry	m	2
		GF basin 1.50 1F basin 0.90		
R11.7 Special pipe fixing taken later.	4/3	Pipes, straight, 40 mm diameter, fixings measured separately	m	2
		1F bath 2.00		
R11.2.3.*		Extra over pipe:		
In this case two sizes of pipe fittings are involved, thus individual items quote the diameter (in contrast to previous extra over pipe items).	4/4	Fittings with two ends 32 mm diam.	nr	4
		Obtuse bent couplings GF basin 2 1F basin 2		
	4/5	Fitting with two ends 40 mm diam.	nr	1
		1F bath 1		
	4/6	Fitting with three ends 40 mm diam. complete with (1) socket plug	nr	1
		1F bath 1		
R11.6.8.1 Also incorporates R11/D2 as trap is jointed to a sanitary appliance.	4/7	Pipework ancillaries, 32 mm diam. plastic bottle traps with 75 mm seal and screw connection to basin waste	nr	2
		GF basin 1 1F basin 1		
	4/8	Pipework ancillaries, 40 mm diam. plastic combination trap with flexible overflow with 75 mm seal and screw connection to bath waste and overflow	nr	1
		1F bath 1		
R11.7	4/9	Fabricated pipe support for 40 mm pipe comprising plastic pipe clip otherwise as item 4/1 above	nr	1
		1F bath 1		

WORKED EXAMPLE 3/1		SHEET NR 5		
Commentary	Item Nr	Description	Unit	Qty
Although this is a separate section in SMM7 it is closely associated with the Disposal Systems and is included here. Note 1 clarifies N13/C1 as discussed earlier in this chapter.		***N13 Sanitary Appliances/Fittings***		
		NOTE:		
		(1) All joints and connections between supply, overflow, waste or soil pipes and the undernoted sanitary appliances/ fittings have been measured with the appropriate pipework.		
Sanitary appliances may be billed in a direct manner quoting maker's catalogue references for each item but in practice a provisional sum is often used to permit the final choice of unit to be left to nearer the installation.		(2) The undernoted sanitary appliances/fittings are included in a Provisional Sum for defined work contained in Section 2 of this bill to cover the supply and delivery to site of these items by a Nominated Supplier.		
'Fixing only' is defined in SMM7 Preliminaries/General Conditions – A52 Nominated Suppliers Rule C1 (SMM7 amendment 2 Feb. 1989)		*Fixing only the following sanitary appliances/fittings:*		
N13.4.1.*.6 Although supplied elsewhere, sufficient information required to judge cost of fixing etc. and risk of damage occurring to unit while in contractor's care	5/1	Low-level close-coupled washdown coloured WC suite comprising vitreous china 'P' trap WC and cistern with seat and lid, fixed to timber floor and masonry wall	nr	1
		GF toilet 1		
	5/2	Low-level siphonic coloured WC suites otherwise as last item	nr	2
		1F bathrooms 2		
	5/3	Lavatory basins 600 × 450 mm comprising vitreous china basin and pedestal, CP plug and chain waste and CP sheerline taps, fixed to masonry wall	nr	2
		1F bathrooms 2		
	5/4	Lavatory basin 500 × 375 mm otherwise as last item	nr	1
		GF toilet 1		
	5/5	Baths 1800 mm long in enamelled pressed steel with plastic panel to one long side, CP plug and chain waste and overflow, CP sheerline taps, fixed to timber floor	nr	2
		1F bathrooms 2		
Sealed into kitchen cabinet by kitchen unit specialist.	5/6	Sink 1310 × 510 mm inset pattern stainless steel with CP plug and chain waste and CP mixer tap set in base by others	nr	1

WORKED EXAMPLE 3/1		SHEET NR 6		
Commentary	Item Nr	Description	Unit	Qty
Overflow pipes are not mentioned in SMM7 but logical to measure under section Y10 and immediately after appliances items.		<u>*Y10 White polythene overflow pipework with solvent welded couplings*</u>		
Y10.1.1.1	6/1	Pipes, straight, 20 mm diameter not supported (short WC overflows)	m	2
Overflows taken through outer walls.		3 WCs 3/0.60		
Y10.2.*		Extra over 20 mm pipe:		
Y10.2.3.3 Pipe fittings not exceeding 65 mm diam. simplified measurement.	6/2	Fittings with two ends	nr	3
		Bent couplings 3		
Y10.2.1 labour on pipe similar in principle to made bends.	6/3	Neat bevelled cut drip ends	nr	3
		3		
Y10.2.2.1	6/4	Special connections to male boss of WC cistern with female plastic connector	nr	3
		3		
		R12 Drainage Below Ground		
R12/P1 – all as D20/P1 specific requirements not covered in general Information provided on Sheet Nr 1.		***Information provided***		
		(i) Ground water was established at 19.550 on (stated date).		
		(ii) Details of the trial pits with their location are given in Specification Clause D20/2.		
		(iii) There are no known live over or underground services affecting this work. An existing land drain will be diverted before this work commences (see Section R13).		
R12/P2		(iv) The layout of the drainage is shown on plan 3/1 Drainage Layout and comprises a separate system for rainwater and foul drainage up to the disconnecting manhole (MH1). Work from MH1 to the sewer is covered by a Provisional Sum contained in Bill Nr 2.		
Work outside curtilage of the site would be kept separate if measured in detail – General Rule 7.1(c)				
R12.1*.*.1 requires commencing levels exceeding 0.25 below existing ground to be stated but in this case it will not affect the drainage work.		(v) Commencing levels stated on Plan 3/1 Drainage Layout are levels after topsoil strip by general contractor.		

WORKED EXAMPLE 3/1			SHEET NR 7	
Commentary	Item Nr	Description	Unit	Qty
SMM7 does not require separation of surface water from foul but is good practice for post-contract adjustments.		***Surface Water Drainage***		
R12.1.1 R12/D1&C1 R12.1*.*.9 backfilling with special materials. Concrete beds/surrounds taken separately as R12.4–6.		*Excavating trenches, pipes not exceeding 200 mm nominal size include backfilling with selected gravel – 75 mm deep below pipes to 75 mm above*		
R12.1.1.2	7/1	Average depth of trench not exceeding 750 mm	m	18
		IC2–RE rear 11.75 2 branches rear 2/1.20 2 branches front 2/1.20 branch yard gully 1.50		
	7/2	Average depth of trench not exceeding 1000 mm	m	22
Net to outer brick face of MH		MH1–IC1 6.00 IC1–IC2 7.50 IC1–RE front 8.00		
R12.3.1	7/3	Disposal, surface water	item	
R12.8.1		*PVCu underground drain pipework with plain ends and polypropylene couplings, in trenches*		
R12.8.1.1	7/4	110 mm diameter	m	41
Pipe taken to inside face of brick MH.		MH1–IC1 6.22 IC1–IC2 7.50		
7.50 m + 0.50 m bend up to surface.		IC1–RE front 8.00		
11.50 m + 0.50 m ditto. Allowing for bends to GL.		IC2–RE rear 12.00 4 branches to RWP 4/1.50 Branch to gully 1.50		
R12.8.1.1.3	7/5	110 mm diameter, vertical	m	1
Rise to GL		4 RWP 4/0.25		
R12.9		Extra over 110 mm pipe:		
R12.9.1.1 Although pipes have plain ends some fittings have sockets (depends on individual manufacturers' designs).	7/6	Single socket bends	nr	3
		Gully 1 2 RE 2		
	7/7	Single socket bends with rests	nr	4
		4 RWP 4		
	7/8	Socketed branches	nr	3
		2 RE 2 Gully 1		

WORKED EXAMPLE 3/1		SHEET NR 8		
Commentary	Item Nr	Description	Unit	Qty
R12.10.1.1 Full value items	8/1	Pipe accessories, 110 mm socketed rodding eyes with alloy screwed grating, set on grade 20 *in situ* concrete foundation 100 mm thick	nr	2
		front & rear 2		
	8/2	Pipe accessories, 110 mm yard gully with 300 mm square top and alloy grating, 550 mm deep with removable bucket, set on grade 20 *in situ* concrete foundation 100 mm thick	nr	1
		front 1		
R12.12.14 – for proprietary units. All inclusive item – no specific mention of excavation but sensible to include it (Rule M9 ambivalent). Inclusions produce effect of what Rule C1 does for trenches.		*(2) PVCu inspection chambers for 110 mm diam. pipes, inclusive of excavation, earthwork support, consolidation, trimming, backfilling and disposal of surplus excavated material*		
	8/3	450 mm Diameter inspection chamber comprising chamber base with blanking plugs set on grade 20 *in situ* concrete foundation 100 mm thick, cast iron cover and frame and chamber risers to suit 820 mm invert	nr	1
		IC2 1		
	8/4	450 mm Diameter inspection chamber all as last item but to suit 910 mm invert	nr	1
		IC1 1		
R12.17.1.*.1&2 Hypothetical specification clause.	8/5	Testing and commissioning the foregoing surface water drainage system as Specification clause R12.44 inclusive of attendance and instrumentation	item	
(Connections to RWPs taken in Section R10)				
		Foul Drainage		
R12.1.1		*Excavating trenches, pipes not exceeding 200 mm nominal size*		
R12.1.1.2	8/6	Average depth of trench not exceeding 1000 mm	m	3
		branch MH2– stack 1.90 branch MH3– stack 2.10		
(Intersections with RW drain ignored)				

WORKED EXAMPLE 3/1		SHEET NR 9		
Commentary	Item Nr	Description	Unit	Qty
R12.1.1.2	9/1	Average depth of trench not exceeding 1250 mm	m	10
		MH1–2 6.50 MH2–3 3.75		
R12.3.1	9/2	Disposal, surface water	item	
R12.4	9/3	Beds to pipes in grade 20 *in situ* concrete 350 mm wide × 100 mm thick	m	13
		Trench ne 1000 3.00 Trench ne 1250 10.25		
R12.8.1		*Vitrified clay underground drain pipework to BS EN 295: 1991 with plain ends and polypropylene couplings, in trenches*		
R12.8.1.1 Measured through to inside face of MH (adding 215 mm at each MH to net trench lengths).	9/4	100 mm diameter	m	16
		MH1–2 6.93 MH2–3 4.18 Branch MH2 2.12 Branch MH3 2.32		
(R12.9 Fittings – NIL) (R12.10 Accessories – NIL) R12.11.* R12/M8 – all work measured in detail		*(3) Brick Manholes*		
D20.2.4.3	9/5	Excavating pits (3 nr), maximum depth not exceeding 2.00 m	m^3	6
		MH1 1.43 × 1.18 × 1.25 MH2 1.18 × 1.18 × 1.25 MH3 1.18 × 1.03 × 1.21		
D20.7.2.1 In theory support not required where trenches occur if dug at same time, but any deductions ignored. Rear face only (see next item).	9/6	Earthwork support maximum depth not exceeding 2.00 m, distance between opposing faces not exceeding 2.00 m	m^2	13
		MH1 1.18 × 1.40 MH2 4/1.18 × 1.25 MH3 2/1.18 × 1.21 2/1.03 × 1.21		
D20.7.2.1.4 D20/D6 affects front and part of sides of MH1 but whole of both sides taken as not worth splitting. Rule explained in Measurement Code with diagram.	9/7	Earthwork support all as last item but next to roadways	m^2	6
		MH1 front 1.18 × 1.40 sides 2/1.43 × 1.40		
D20.8.1	9/8	Disposal, surface water	item	
D20.8.3.1 No backfilling required.	9/9	Disposal, excavated material, off site	m^3	6
		Vol. item 9/5 above = 5.57 m^3		

WORKED EXAMPLE 3/1			SHEET NR 10	
Commentary	Item Nr	Description	Unit	Qty
D20.13.2.3	10/1	Surface treatments, compacting bottoms of excavations	m^2	4
		MH1 1.43 × 1.18 MH2 1.18 × 1.18 MH3 1.18 × 1.03		
E10.3.0.5	10/2	*In situ* concrete bed of grade 20 concrete, 20 mm aggregate, thickness not exceeding 150 mm, poured on or against earth	m^3	1
		Area of item 10/1 = 4.30 m^2 × 0.15		
F10/S1		*Brickwork in engineering bricks to BS 3921 Class B built in cement mortar 1:3 in English bond, flush pointed as work proceeds*		
F10.1.2.1 F10/D3 *Waste calculations:* MH nr MH1 MH2 MH3 Total depth = 1.55 1.35 1.31 Dt base 0.15 slab/cvr 0.20 = 0.35 0.35 0.35 Net bwk ht = 1.20 1.00 0.96 Centre line lengths of brickwork.	10/3	Walls, one brick thick, pointed one side MH1 2/1.22 × 1.20 2/0.97 × 1.20 MH2 2/0.97 × 1.00 2/0.97 × 1.00 MH3 2/0.97 × 0.96 2/0.82 × 0.96	m^2	13
R12.11.7.1 C6	10/4	Building in ends of pipes into one brick thick walls of manholes; 100 mm clay pipes	nr	6
		MH1 1 MH2 3 MH3 2		
	10/5	Ditto; 110 mm PVCu pipe	nr	1
(Assume hole for exit pipe from MH1 will be covered by Prov. Sum.) R12.11.8.1		MH1 1		
		Channels in vitrified clay to BS EN 295: 1991, half round with spigot joints jointed in cement mortar 1:3, set in separately measured concrete benching		
	10/6	100 mm diameter, curved girth 600 mm	nr	1
		MH3 1		
	10/7	100 mm diameter, 750 mm long with 1 branch intersection 350 mm long	nr	1
		MH2 1		
	10/8	100 mm diameter, 1000 mm long with 1 branch intersection curved girth 600 mm to suit 250 mm higher invert	nr	1
		MH3 1		

WORKED EXAMPLE 3/1		SHEET NR 11		
Commentary	Item Nr	Description	Unit	Qty
R12.11.9 Could well be combined with channels, each MH having one item covering both elements.		*Benching of* in situ *concrete grade 20, 20 mm aggregate, average 150 mm thick and floated smooth to falls to channels*		
R12.11.9.1	11/1	To manhole 750 × 600 mm MH3 1	nr	1
	11/2	To manhole 750 × 750 mm MH2 1	nr	1
	11/3	To manhole 1000 × 750 mm	nr	1
R12.11.13.1		*Manhole intake cover slabs in precast concrete, grade 25, reinforced with 12 mm diam. MS rods to BS 4449 at 100 mm centres, each fabricated in (4) pieces with recessed joints to provide 600 × 450 mm opening*		
	11/4	To manhole 1180 × 1030 mm overall MH3 1	nr	1
	11/5	To manhole 1180 × 1180 mm overall MH2 1	nr	1
	11/6	To manhole 1430 × 1180 mm overall MH1 1	nr	1
R12.11.11.1	11/7	Manhole covers and frames in ductile iron to BS 497 Grade 3, 600 × 450 mm, frame set in cement mortar 1:3 and cover set in grease MH1,2,3 3	nr	3
End of manholes				
R12.17.1.2.1 & 2	11/8	Testing and commissioning the foregoing foul drainage system as Specification Clauses R12.45–46, inclusive of attendance and instrumentation	item	

WORKED EXAMPLE 3/1		SHEET NR 12		
Commentary	Item Nr	Description	Unit	Qty
		R13 Land Drainage		
R12/P1 – all as D20/P1 specific requirements not covered in general Information provided on Sheet Nr 1.		***Information provided***		
		(i) Ground water was established at 19.550 on (stated date).		
		(ii) Details of the trial pits with their location are given in Specification Clause D20/2.		
		(iii) There are no known live over or underground services affecting this work.		
		(iv) The layout of the land drainage is shown on Plan 3/1 Drainage Layout and comprises diverting an existing land drain around the new house.		
R13.1.1 R13/C1 R13.1*.*.9 backfilling with special materials.		*Excavating trenches, pipes not exceeding 200 mm nominal size include backfilling with graded gravel – 75 mm deep below pipes to 100 mm above*		
R13.1.1.2	12/1	Average depth of trench not exceeding 500 mm	m	13
Scaled from plan		7.50 5.00		
R13.3.1	12/2	Disposal, surface water	item	
R13.8.1		*PVCu underground perforated flexible drain pipework, in trenches*		
R13.8.1.1	12/3	90 mm diameter	m	13
		Qty. as trench item 12/1=12.50		
Additional rules – work to existing buildings R13 Drainage (SMM7 – page 173). Rule 1.1.1.4 combined with rule 3.1 provides a clear enumerated item.	12/4	Break into existing porous tile land drain and joint to new 90 mm flexible pipe at diversion, include sealing off old onward pipe with *in situ* concrete plug	nr	1
		West 1		
	12/5	Break into existing porous tile land drain and joint to new 90 mm flexible pipe at end of diversion	nr	1
		South 1		

4 Measurement of Service Plumbing Work

Introduction

Service plumbing work comprises the provision of mains water service into buildings and the distribution of hot and cold water services within buildings. In Chapter 2 the general approach to billing and measurement of building services was covered, indicating three basic components for supply services installations – namely: source, distribution and outlets. Examples of the respective components for service plumbing are:

Source	**Distribution**	**Outlets**
Connections to main	Distribution pipework	Connections to appliances
Cold water storage		
Hot water storage		

Measurement Rules

Service plumbing is covered by SMM7 Appendix B/Section S 'Piped supply systems' for the classification of the installation and Rules Y10–59 for the detailed rules of measurement.

'**Source**' items are mainly enumerated under SMM7 Rules Y20–25 General pipeline equipment. These are defined in the Measurement Code (page 42) which lists such items as tanks, cylinders and calorifiers. The actual classifications within Rules Y20–25 are listed in the Detailed Contents of SMM7 on page 7, and these references are used in the Commentary columns of the Worked Examples which follow.

The connection to the public main water supply is enumerated as extra over the pipe in which it occurs under SMM7 Rule Y10.2.2 Special joints and connections as defined in Rule D2.

'**Distribution**' items are invariably pipelines which are measured under SMM7 Rules Y10/11 Pipelines and Pipeline ancillaries. This generally involves pipes being measured by length, with fittings and labours enumerated as 'extra over the pipes in which they occur' as Y10.1 and Y10.2. Examples of pipe fittings which are 'extra over' are given in the

Measurement Code (page 41) and include such items as bends, junctions and tees. Pipework ancillaries, on the other hand, are enumerated *full value* as Y11.8, and these are also defined in the Measurement Code as such items as valves and stop cocks. It is obviously necessary to differentiate between fittings and ancillaries in order to itemise them correctly but in practice, on a practical level, there is rarely any need to deduct the length of pipe displaced by any full-value ancillary. For example, a valve incorporated in normal sized pipework will only cause a saving of about 50–100 mm of pipe, which is minimal when compared with the practical level of accuracy of measuring the pipe from a drawing in the first place. In the case of larger diameters of service pipes, such deductions could however become significant.

'**Outlets**' are confined to the pipe connections to appliances, as the sanitary appliances/fittings themselves have been measured as 'inlets' of Disposal Systems in Chapter 3. The rationale behind the measurement of connections to such appliances/fittings is fully explained in Chapter 3. Such connections are dealt with by SMM7 Rule Y10.2.2 Special joints and connections as defined in Rule D2 and measured as 'extra over the pipes in which they occur'.

Taking Off Pipelines

The task of taking off the quantities of service pipelines along with their associated labours, fittings, connections and ancillaries is most efficiently tackled by adopting the abstract sheet technique, as explained with examples in Chapter 2. The worked examples in this chapter adopt this approach.

WORKED EXAMPLE

The following **Worked Example 4/1** will demonstrate the application of the various Rules from SMM7 to typical service plumbing installations. Readers should refer to SMM7 and the Measurement Code while following this example, which will be cross-referenced where appropriate. Worked example 4/1 is based on the same design of detached two-storey house as used in Chapter 3 and incorporates mains water, cold and hot water installations. The cold water section is not measured in detail, as it is repetitious and would not introduce any additional items to those sections included in full.

This worked example is set out in draft bill format, which clarifies not only the SMM7 measurement requirements but also various ancillary matters such as testing and sundries which may not appear clearly in a basic take-off.

Worked Example 4/1: Detached House – Service Plumbing Work

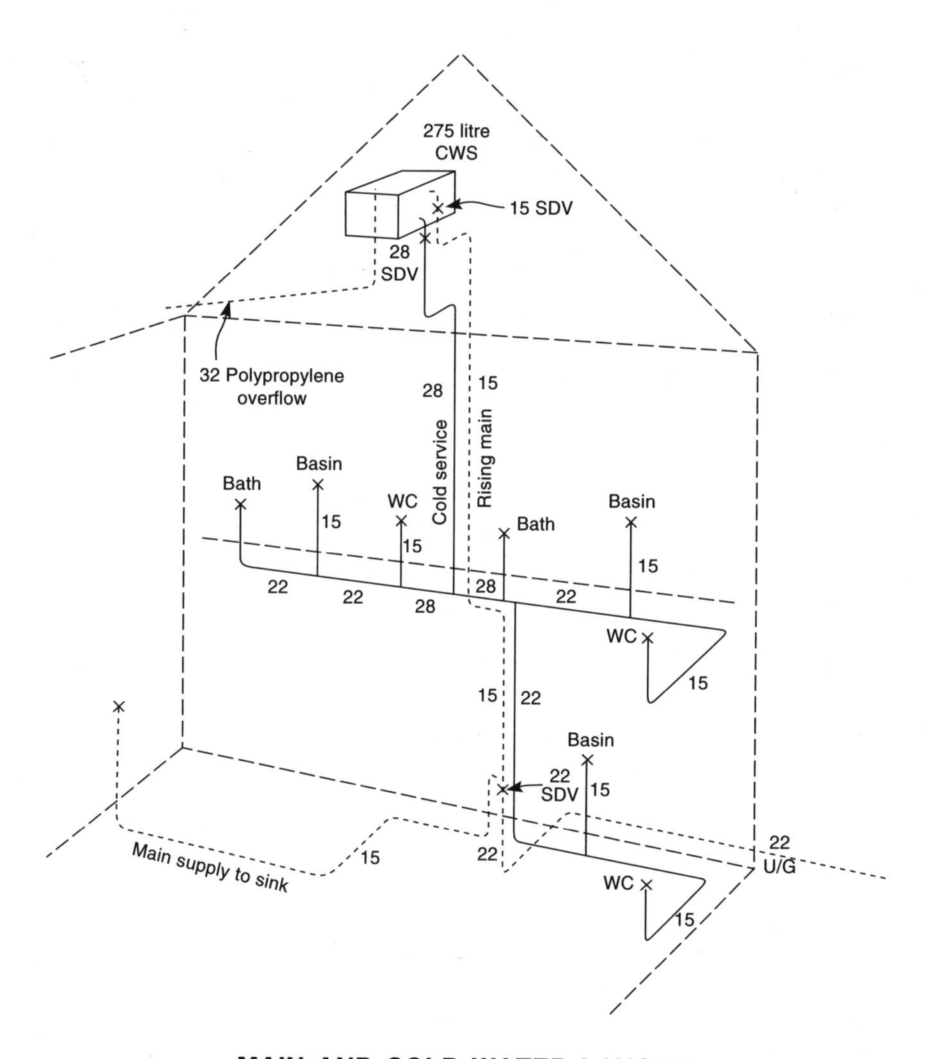

MAIN AND COLD WATER LAYOUT

(Refer to Plans 3/1/1–3 for further details of detached house)

4/1/1

Worked Example 4/1: Detached House – Service Plumbing Work

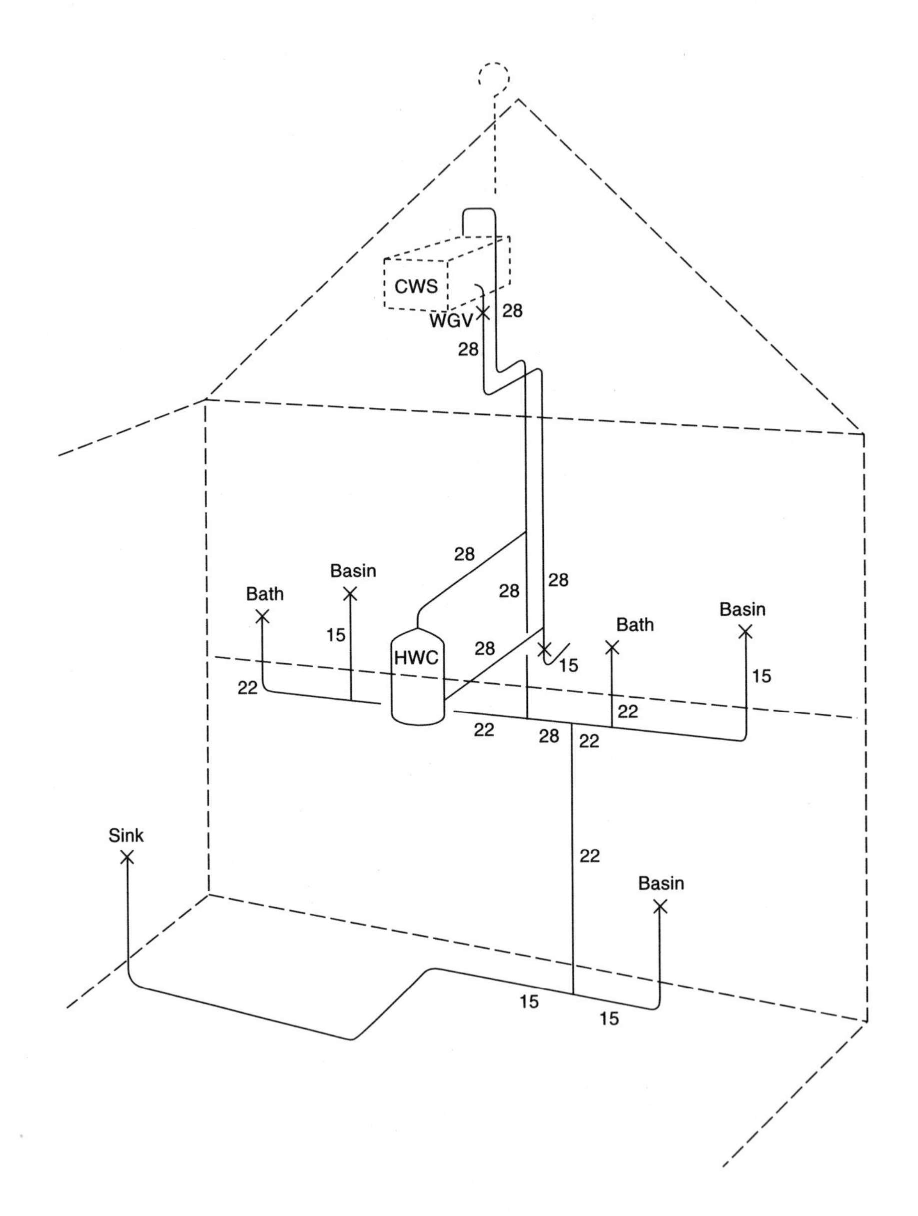

MAIN AND COLD WATER LAYOUT

(Refer to Plans 3/1/1–3 for further details of detached house)

4/1/2

WORKED EXAMPLE 4/1 – DETACHED HOUSE
Abstract Sheets for SMM7 Section S – Piped Supply Systems

Main Supply
Note: 22 mm Underground main billed direct without an abstract as straightforward take off.

ABSTRACT NR 1 – Main supply – 15 mm copper with saddle clips; compression fittings; timber background

Location	Pipe	Made Bends	Made Offsets	Fittings 2 ends	Special Connections female bent	SDV
Roofspace	2.60	1	1	2	1	1
Under GF	4.50	2	1	1		
1F	1.40	2		2		
Totals	8.50	5	2	5	1	1

ABSTRACT NR 2 – Main supply – copper with pipe rings; compression fittings; masonry background

Location	Pipe		Made Bends 15	Made Offsets 15	Fittings 2 ends 15	Fittings 3 ends 22	Special Connections female straight 15
	22	15					
GF Toilet	0.25	2.40	2	1		1	
GF Branch		0.40	2		1		
1F Cupb.		2.50	2	1			
GF Sink		1.00		2			1
Totals	0.25	6.30	6	4	1	1	1

Commentary: Abstract Nr 1 is only concerned with one size of pipe and thus sizes can be omitted from table.
The 'Fittings 2 ends' are bent couplings and the 'Fitting 3 ends' is a tee piece 22 × 15 × 15 mm.

Cold Supply
Note: The cold supply installation comprises pipework only which closely follows the equivalent items in the Main Supply and Hot Water sections. It has therefore been omitted to avoid repetitious examples. This work would have been contained in Abstracts Nrs 3–5.

Hot Water Installation

ABSTRACT NR 6 – Hot water – copper with saddle clips; compression fittings; timber background

Location	Pipe			Made Bends			Made Offsets			Fittings 2 ends			Fittings 3 ends		Special Connections female bent 28	WGV 28
	28	22	15	28	22	15	28	22	15	28	22	15	28	22		
Roofspace																
cold feed	2.20			2			1			2					1	1
1F floor		0.50	3.20	1.50		2	1	1	2	1		1	1	2	2	
Totals	2.70	3.20	1.50	2	2	1	2	2	1	2	1	1	2	2	1	1

ABSTRACT NR 7 – Hot water – copper with pipe rings; compression fittings; masonry background

Location	Pipe			Made Bends			Made Offsets			Fittings 2 ends		Fittings 3 ends		Special Connections			Drain Down Valve 15
	28	22	15	28	22	15	28	22	15	28	15	28	22	male straight 28	male bent 28	bent hose 15	
1F cupb.																	
cold feed	4.30			1			2					1		1			
Expansion	5.00			1			1					1			1		
Drain down			0.60			1										1	1
Roofspace	3.00			2			1			2							
GF drop		2.70			1			1					1				
GF floor			4.50			2			2		4						
Totals	12.30	2.70	5.10	4	1	3	4	1	2	2	4	2	1	1	1	1	1

Hot Water Installation (cont.)
Note: The specification requires all pipes within bathrooms, toilets and kitchen to have capillary fittings, which have a better appearance if exposed.

ABSTRACT NR 8 – Hot water – copper with pipe rings; capillary fittings; masonry background

Location	Pipe		Made Bends		Made Offsets		Fittings 2 ends 15	Special Connections female straight	
	22	15	22	15	22	15		22	15
1F Bathroom	0.60	0.90	1	1	1	1	1	1	1
1F En-suite	0.60	0.90	1	1	1	1	1	1	1
GF Toilet		0.90		1		1	1		1
GF Kitchen		1.00				1			1
Totals	1.20	3.70	2	3	2	4	3	2	4

WORKED EXAMPLE 4/1		SHEET NR 1		
Commentary	Item Nr	Description	Unit	Qty
		DETACHED HOUSE		
Sequential order of bills		*BILL NR 9*		
Common Arrangement of Work Section heading		*S – PIPED SUPPLY SYSTEMS*		
		Information provided		
This note covers the requirements of Rule P1 of sections Y10–59. Placing this first removes the need for repetition within the various subsections of the bill.		The following represents the installation of mains, cold and hot water service plumbing including sundries and thermal insulation as detailed on Plans 4/1/1 and 4/1/2.		
Heading from Appendix B – but further categorised for mains and location (helpful post-contract information).		***S10 Cold Water Installation – Main Supply – Underground***		
Essential heading to indicate the rules adopted for the measurements following.		***Y 10/11 Pipelines and Pipeline Ancillaries***		
Table Y – fully annealed copper tube supplied in coils up to 20 metres.		*Copper tubing to BS 2871: Part 1 – Table Y*		
Y10.1.3.1.3 Manipulative compression fittings necessary for this tubing and pressure.	1/1	Pipes, flexible, 22 mm diameter jointed with brass manipulative compression couplings, laid in trenches	m	13
Straightforward take-off; no abstract necessary.		water authority stop valve – house wall 12.50		
Y10.1.3.1.2	1/2	Pipes, flexible, 22 mm diameter jointed as above, in ducts	m	1
Fireclay duct by builder.		substructure 0.90		
Y10.1.3.1.1	1/3	Pipes, flexible, 22 mm diameter jointed as above, fixed to masonry background with brass pipe rings plugged and screwed at centres as specified	m	1
		up to GF 0.95		
Y10.2.2.1 + General Rule 7.1(c) Work outside the curtilage of site. About 300 mm of pipe from item 1/1 would also be in the public footpath but this is ignored as not cost significant.	1/4	Extra over copper pipes for special connection to Water Authority's stop valve in public footpath comprising 22 mm diameter brass female straight connection with manipulative compression joint to pipe	nr	1
		just within footpath 1		
Y11.8 + Measurement Code Y11/C7	1/5	Pipework ancillary, 22 mm diameter brass screw down valve with manipulative compression joints	nr	1
		just above GF 1		

WORKED EXAMPLE 4/1			SHEET NR 2	
Commentary	Item Nr	Description	Unit	Qty
Heading from Appendix B with additional information.		***S10 Cold Water Installation – Main Supply – Internal***		
Essential heading to indicate rules adopted for measurements following.		***Y10/11 Pipelines and Pipeline Ancillaries***		
Table X – normal quality copper tube.		*Copper tubing to BS 2871: Part 1 – Table X*		
Y10.1.1.1.1 Non-manipulative compression joints assumed specified above GL would be clarified in Specification.	2/1	Pipes, straight, 15 mm diameter jointed with brass compression couplings, fixed to timber background with copper saddle clips at centres as specified	m	9
Two stage take-off by abstract – see explanation in Chapter 2.		Abstract nr 1 8.50		
Y10.1.1.1.1	2/2	Pipes, straight, 15 mm diameter jointed as above, fixed to masonry background with brass pipe rings plugged and screwed at centres as specified	m	6
		Abstract nr 2 6.30		
General Rule 3.3 – small quantities given as one unit. Short length between SDV and tee, not required to be fixed.	2/3	Pipes, straight, 22 mm diameter jointed as above, not fixed	m	1
		Abstract nr 2 0.25		
Y10.2		*Items extra over copper pipes in which they occur*		
Y10.2.1	2/4	Made bends 15 mm pipe	nr	11
		Abstract nr 1 5 Abstract nr 2 6		
Not in SMM but justified as more labour content than 1 bend but less than 2 bends.	2/5	Made offsets 15 mm pipe	nr	6
		Abstract nr 1 2 Abstract nr 2 4		
Y10.2.3.3.2 Copper pipe fittings require to be identified owing to the many patterns and metals specified.	2/6	Fittings 15 mm diameter, brass compression pattern two ends	nr	6
		Abstract nr 1 5 Abstract nr 2 1		
Y10.2.3.4.2	2/7	Fittings 22 mm diameter, brass compression pattern, three ends	nr	1
		Abstract nr 2 1		
Y10.2.2.1	2/8	Special connection between 15 mm pipe and male threaded equipment with brass compression female straight connection	nr	1
		Abstract nr 2 1		

WORKED EXAMPLE 4/1			SHEET NR 3	
Commentary	Item Nr	Description	Unit	Qty
Extra over items continued.	3/1	Special joint between 15 mm pipe and male threaded equipment with brass compression female bent connection	nr	1
		Abstract nr 1 1		
Y11.8 + Measurement Code Full value – heading prevents 'extra over' continuing.		*Pipework Ancillaries*		
	3/2	Brass screw down valve 15 mm diameter with compression joints to copper pipes	nr	1
		Y20/25 General Pipeline Equipment		
Y21.1.1.1.1 Measurement Code	3/3	Cold water storage cistern comprising 275 litre actual capacity plastic cistern make and model as specified complete with plastic lid, holed for and provided with: 15 mm diameter brass ballvalve with male tail, jam nuts and large backplate, 350 mm brass shank and 175 mm plastic ball; 32 mm diameter plastic standing waste overflow with plastic screwed outlet; (2) 28 mm diameter brass male outlet bosses with jam nuts; support shelf by others	nr	1
		roofspace 1		
Overflow pipes are not mentioned in SMM7 but logical to measure under section Y10 and immediately after equipment.		*Y10 polypropylene overflow pipework jointed with push fit 'O' ring couplings*		
Y10.1.1.1.1	3/4	Pipes, straight, 32 mm diameter fixed to timber background with plastic pipe clips screwed at centres as specified	m	5
		roofspace 4.60		
Y10.2		*Items extra over polypropylene pipes in which they occur*		
Not in SMM7 but labour similar in principle to Y10.2.1.	3/5	Neat cut bevelled end 32 mm pipe	nr	1
		eaves 1		
Y10.2.3.3.2	3/6	Fittings 32 mm diameter, 'O' ring pattern, two ends	nr	1
		elbow 1		
Y10.2.2.1	3/7	Special connection between 32 mm pipe and male threaded equipment with 'O' ring female straight connection	nr	1
		to cistern 1		

WORKED EXAMPLE 4/1			SHEET NR 4	
Commentary	Item Nr	Description	Unit	Qty
Heading from Appendix B with additional information.		***S10 Cold Water Installation – Cold Supply***		
Essential heading to indicate rules adopted for measurements following. Table X – normal quality copper tube.		***Y10/11 Pipelines and Pipeline Ancillaries*** <u>*Copper tubing to BS 2871: Part I – Table X*</u>		
Here would follow the detailed measurement of the cold supply pipework which closely follows the equivalent items in the Main Supply and the Hot Water sections. There is no requirement for Equipment in this section – thus it is more appropriate to detail the measurement of the Hot Water Installation.				
Heading from Appendix B.		***S11 Hot Water Installation***		
Essential heading to indicate rules adopted for measurements following Table X – normal quality copper tube.		***Y10/11 Pipelines and Pipeline Ancillaries*** <u>*Copper tubing to BS 2871: Part 1 – Table X*</u>		
Two-stage take-off by abstract – see explanation in Chapter 2. Y10.1.1.1.1		Pipes, straight, jointed with brass compression couplings, fixed to timber background with copper saddle clips at centres as specified		
Different diameters in sub-items.	4/1	15 mm diameter	m	2
		Abstract nr 6 1.50		
	4/2	22 mm diameter	m	3
		Abstract nr 6 3.20		
	4/3	28 mm diameter	m	3
		Abstract nr 6 2.70		
Different background.		Pipes, straight, jointed with brass compression couplings, fixed to masonry background with brass pipe rings plugged and screwed at centres as specified		
	4/4	15 mm diameter	m	5
		Abstract nr 7 5.10		
	4/5	22 mm diameter	m	3
		Abstract nr 7 2.70		
	4/6	28 mm diameter	m	12
		Abstract nr 7 12.30		

WORKED EXAMPLE 4/1			SHEET NR 5	
Commentary	Item Nr	Description	Unit	Qty
Different type of coupling. Capillary fittings for potable water must have non-lead-based solder.		Pipes, straight, jointed with copper capillary couplings, fixed to masonry background with brass pipe rings plugged and screwed at centres as specified		
Y10.1.1.1.1	5/1	15 mm diameter	m	4
		Abstract nr 8 3.70		
	5/2	22 mm diameter	m	1
		Abstract nr 8 1.20		
Y10.2		*Items extra over copper pipes in which they occur*		
Y10.2.1		Made bends		
Different pipe sizes in sub-items. Items indented once for 'extra over' and once more for sub-items.	5/3	15 mm pipe	nr	7
		Abstract nr 6 1 Abstract nr 7 3 Abstract nr 8 3		
	5/4	22 mm pipe	nr	5
		Abstract nr 6 2 Abstract nr 7 1 Abstract nr 8 2		
	5/5	28 mm pipe	nr	6
		Abstract nr 6 2 Abstract nr 7 4		
Not in SMM but justified as more labour content than 1 bend but less than 2 bends.		Made offsets		
	5/6	15 mm pipe	nr	7
		Abstract nr 6 1 Abstract nr 7 2 Abstract nr 8 4		
	5/7	22 mm pipe	nr	5
		Abstract nr 6 2 Abstract nr 7 1 Abstract nr 8 2		
	5/8	28 mm pipe	nr	6
		Abstract nr 6 2 Abstract nr 7 4		
In Scottish plumbing practice the hot water expansion would be taken out through the roof to terminate in a swan-neck bend. This would also be measured as Y10.2.1 (adapted). Lead slates for same as H71.26.				

WORKED EXAMPLE 4/1		SHEET NR 6		
Commentary	Item Nr	Description	Unit	Qty
Important heading repeated.		*Items extra over copper pipes in which they occur (continued)*		
Y10.2.3.3.2 – Compression type kept separate from capillary.		Fittings, brass compression pattern, two ends		
	6/1	15 mm pipe	nr	5
		Abstract nr 6 1 Abstract nr 7 4		
	6/2	22 mm pipe	nr	1
		Abstract nr 6 1		
	6/3	28 mm pipe	nr	4
		Abstract nr 6 2 Abstract nr 7 2		
Y10.2.3.4.2		Fittings, brass compression pattern, three ends		
	6/4	22 mm pipe	nr	3
		Abstract nr 6 2 Abstract nr 7 1		
	6/5	28 mm pipe	nr	4
		Abstract nr 6 2 Abstract nr 7 2		
Y10.2.3.3.2 – Capillary type and in this project only one size required.	6/6	Fittings, copper capillary pattern, two ends; 15 mm pipe	nr	3
		Abstract nr 8 3		
Y10.2.2.1	6/7	Special connection between 28 mm pipe and female threaded equipment with brass compression male straight connection	nr	1
		Abstract nr 7 1		
	6/8	Special connection between 28 mm pipe and male threaded equipment with brass compression female bent connection	nr	1
		Abstract nr 6 1		
	6/9	Special connection between 28 mm pipe and female threaded equipment with brass compression male bent connection	nr	1
		Abstract nr 7 1		

WORKED EXAMPLE 4/1			SHEET NR 7	
Commentary	Item Nr	Description	Unit	Qty
Important heading repeated.		*Items extra over copper pipes in which they occur (continued)*		
Y10.2.2.1 – drain down at first floor level. Connection allows for future maintenance.	7/1	Special connection between 15 mm pipe and future Client supplied hose coupling with brass compression bent threaded hose connection	nr	1
		Abstract nr 7 1		
Y10.2.2.1 – two sizes of same connection so sub-item style adopted.		Special connections between pipe and male threaded equipment with copper capillary female straight connection		
	7/2	15 mm pipe	nr	4
		Abstract nr 8 4		
	7/3	22 mm pipe	nr	2
		Abstract nr 8 2		
Y11.8 + Measurement Code Full value – heading prevents 'extra over' continuing.		*Pipework Ancillaries*		
	7/4	Brass drain down cock 15 mm diameter with compression joints to copper pipes	nr	1
		Abstract nr 7 1		
	7/5	Brass wheel gate valve 28 mm diameter with compression joints to copper pipes	nr	1
		Abstract nr 6 1		
		Y20/25 General Pipeline Equipment		
Y23.1.1.1.4	7/6	Copper hot water storage cylinder, indirect 135 litre capacity to BS 699 (Grade 2) as specified, with immersion heater boss, (2) 28 mm female threaded primary bosses and (2) 28 mm female threaded secondary bosses precoated with 32 mm thick foam insulation; support shelf by others	nr	1
		1F cupb. 1		
Y23.2.1.1	7/7	Ancillary for equipment supplied by others, take delivery of and fix immersion heater to last item	nr	1
		Electrician 1		

WORKED EXAMPLE 4/1			SHEET NR 8	
Commentary	**Item Nr**	**Description**	**Unit**	**Qty**
In large projects thermal insulation could be separately measured in each subsection of the bill.		***Y50 Thermal Insulation***		
Y50.1.1.1		*Plastic faced glass fibre sectional insulation minimum 15 mm thick, secured with waterproof adhesive tape*		
	8/1	Pipelines, 15 mm diameter	m	*
	8/2	Pipelines, 22 mm diameter	m	*
	8/3	Pipelines, 28 mm diameter	m	*
* These quantities would readily be identified from Abstracts 1–8 where pipes are in ducts, roofspace, below floors etc.				
Proprietary insulation.		*Insulation packs in preformed foam as specification clause Y50/5 manufactured by Insulpak plc*		
Y50.1.4.2	8/4	Equipment, cold water storage cistern, nominal size 900 × 650 × 630 mm to (4) sides and lid in 40 mm thick foam, manufacturer's reference nr CWS/966	nr	1
		roofspace 1		
Sensible to group these items for whole of Piped Supply Systems.		***Y51–59 Testing and Sundries***		
Y59.1.1	8/5	Marking positions of holes, mortices and chases in the structure for the whole of the foregoing Piped Supply System	item	
Testing and commissioning of this type of installation would only comprise simple requirements. Y51.4.1	8/6	Testing and commissioning the whole of the foregoing Piped Supply System as per specification clause Y51/1	item	
P30 Rule M2		***P30/31 Trenches, Holes, Chases for Piped Supply System***		
P30 Information Provided not covered at start of this Bill – therefore required here. As full details of ground water levels, trial pits etc. are in the Drainage Section it is enough to cross-reference in this case. (See Chapter 3 – Worked Example 3/1.)		***Information Provided***: for information regarding the nature of excavation work refer to the Bill Nr 8 – Disposal Systems, Section R12 Drainage Below Ground.		
P30.1.1.2	8/7	Excavating trenches to receive pipe not exceeding 200 mm nominal size include bedding in sand at least 100 mm thick all round pipe: average depth not exceeding 500 mm	m	12
P30 Rules M3, M4, D2, C1		whole route (scaled) 12.00		

WORKED EXAMPLE 4/1			SHEET NR 9	
Commentary	Item Nr	Description	Unit	Qty
General Rule 7.1(c) work outside the boundary of the site.	9/1	Excavating trenches all as last item but work outside the curtilage of the site	m	1
General Rule 3.3 minimum quantity.		In public footpath 0.45 Main water connection		
P30.2		*Items extra over excavating trenches, irrespective of depth*		
P30.2.1.1 – Likely occurrence hence inclusion of assumed quantity.	9/2	Breaking out existing rock, PROVISIONAL	m³	1
P30–M6		Assumed 4.00 × 0.50 × 0.30		
P30.2.2.5.1 Width determined by Rule Y30–M6 minimum 500 mm in this case. General Rule 7.1(c)	9/3	Breaking out existing public footpath comprising tarmacadam on bottoming and neatly reinstate to match existing to specification clause Q25.18, work outside the curtilage of the site	m²	1
		Main water connection 0.45 × 0.50		
Line to cancel 'extra over'		- - - - - - - - - - - - - -		
P30.3.1 Always given as risk item	9/4	Disposal, surface water	item	
P30.13.1.1	9/5	Identification tape, straight, PVC strip incorporated into trench backfilling 300 mm above main water pipe	m	12
		Item 8/7 12.00 Item 9/1 0.45		
P31.20 Cutting or forming holes for services installations are enumerated, see Chapter 2 – 'Builder's Work'.	9/6	Cutting or forming holes for pipes not exceeding 55 mm nominal size through concrete blockwork partitions 75 mm thick and making good	nr	2
		Hot feed + expan. 1F 2		
	9/7	Cutting or forming hole for pipe as last item but through 300 mm thick cavity walling comprising concrete blockwork inner leaf and facing brick outer leaf and making good	nr	1
		Hot drain down 1F 1		
Group of pipes taken as special item as cheaper to form than 3 individual holes.	9/8	Cutting or forming hole for group of 3 pipes each not exceeding 55 mm nominal size through concrete blockwork partition 100 mm thick and making good	nr	1
		Main, hot & cold GF ceiling 1		

5 Measurement of Heating Installations

Introduction

Space heating within buildings can be achieved by three main methods, namely:

(1) Hot water in pipes as the transfer medium between the heat source and the heat-emitting equipment.
(2) Warmed air in ducts as the transfer medium between the heat source and the warm air outlets.
(3) Electrical heating which comprises electrical wiring and electrical heating appliances emitting heat at the required location.

As these three methods involve completely different technologies and installation techniques they are separately measured under SMM7. This chapter will cover those installations based on hot water while the warm-air-based systems are covered in Chapter 6, and electrical heating is covered in Chapter 7.

Hot water heating systems can vary from micro bore/small bore systems for domestic property up to very large installations for public and commercial buildings. The smaller systems are frequently executed in copper tubing and their measurement is therefore similar to that of service plumbing, as discussed in Chapter 4. Medium to large systems, however, are usually installed using the more economical mild steel tubing with screwed connections, which requires an understanding of the particular installation techniques in order to measure this work successfully.

In Chapter 2 the general approach to billing and measurement of building services was covered, indicating three basic components for supply services – source, distribution and outlets. Examples of the respective components for heating installations are:

Source	**Distribution**	**Outlets**
Boiler plant	Distribution pipework	Radiators
Connections to heating mains		Heating batteries
Calorifiers		Unit heaters

Measurement Rules

Heating installations (based on hot water) are covered by SMM7 Appendix B/Section T Mechanical heating/Cooling/Refrigeration systems for the classification of the installation, and Rules Y10–59 for the detailed rules of measurement.

'**Source**' items are mainly enumerated under SMM7 Rules Y20–25 General pipeline equipment. These are defined in the Measurement Code (page 42) which lists items such as boilers, fuel handling units and calorifiers. The actual classifications within Rules Y20–25 are listed in the Detailed Contents of SMM7 on page 7, and these references are used in the Commentary columns of the Worked Examples which follow.

Should there be a connection to a heating main, then this would be enumerated as extra over the pipe in which it occurs under SMM7 Rule Y10.2.2 Special joints and connections, and defined in Rule D2.

'**Distribution**' items are invariably pipelines which are measured under SMM7 Rules Y10/11 Pipelines and Pipeline ancillaries. This generally involves pipes being measured by length with fittings, special joints and labours enumerated as 'extra over the pipes in which they occur' as Y10.1 and Y10.2. Examples of pipe fittings which are 'extra over' are given in the Measurement Code (page 41) and include such items as bends, junctions, tees and unions. Pipework ancillaries on the other hand are enumerated *full value* as Y11.8, and these are also defined in the Measurement Code as such items as valves, mixing valves, cocks and gauges. It is obviously necessary to differentiate between fittings and ancillaries in order to itemise them correctly, but on a practical level there is only the need to deduct the length of pipe displaced by a full value ancillary on larger diameter pipes. For example, a valve incorporated in copper or steel pipework up to 35/40 mm diameter will only cause a saving of about 50–100 mm of pipe, which is minimal when compared with the practical level of accuracy of measuring the pipe from a drawing in the first place. In the case of larger diameters of heating pipes, such deductions could however become significant.

It should be noted that radiator valves are *not* classed as pipework ancillaries but as ancillaries to equipment – refer to the paragraph on 'Outlets' below.

Pumps

Pumps are frequently incorporated in pipe runs but are defined in the Measurement Code as Y20–25 General pipeline equipment and should be enumerated *full value* as Y20–25.1, which has the same effect as if they were pipework ancillaries. Pumps do replace sizable lengths of pipe and are often flanked with isolating valves, thus meriting an appropriate deduction from the pipe.

Demountable Couplings
Mild steel screwed pipes to BS 1387, by necessity of installation and future maintenance, require to have incorporated demountable couplings or demountable elbows at appropriate locations throughout installations. Normally the services design engineer should indicate these fittings on the plans, but if not the surveyor will still require to include for them in the bill. Demountable joints are measured as Special joints and connections under Y10.2.2 as 'extra over the pipes in which they occur'. They are defined under Rule D2 as joints which differ from those generally occurring in the running length. This requirement complements Rule C3, which deems pipes to include joints in their running length that in the case of screwed steel pipe would be plain screwed sockets. Such pipes are normally sold in 6 m lengths each with a socket supplied.

'**Outlets**' comprise heating output appliances of various types which are enumerated under SMM7 Rules Y20–25 General pipeline equipment. These are defined in the Measurement Code (page 42) which lists items such as heaters, heating batteries, radiators and convectors. The actual classifications within Rules Y20–25 are listed in the Detailed Contents of SMM7 on page 7, and these references are used in the Commentary columns of the Worked Examples which follow.

Control valves for such pieces of equipment are enumerated as 'Ancillaries for equipment not provided with the equipment' Rules Y20–25.2 and defined in the Measurement Code (page 42) which would include radiator valves, thermostatic radiator valves and lockshield valves.

Special Locations of Work

Special locations of pipes such as in ducts, trenches, chases, floor screeds and *in situ* concrete are a requirement of Y10.1*.*.2–6. In addition, all work in plant rooms is required to be identified separately by Rule M2 of Sections Y10/11, Y20–25 and Y50. Plant rooms are defined in the Measurement Code as including heating chambers, ventilation machinery rooms, tank roofs, etc. However the commonest plant rooms are likely to be boiler rooms and calorifier rooms.

The cost significance of these special locations is generally one of additional labour costs due to lack of working space and/or specific sequencing requirements. Items for pipes are readily qualified with the required additional descriptions while any work within plant rooms can be dealt with similarly, or if more appropriate under a separate heading grouping all such work together.

Taking Off Pipelines

The task of taking off the quantities of heating pipelines along with their associated labours, fittings, connections and ancillaries is most efficiently tackled by adopting the abstract sheet technique as explained with examples in Chapter 2. The worked examples in this chapter adopt this approach.

Boiler Flues and Chimneys

Neither flues nor chimneys feature in the Alphabetical Index of SMM7 but they are covered in Section Y20–25 General pipeline equipment – Rule 7 Independent vertical steel chimneys, and associated Measurement Rule M3 which deals with flue pipes. The basis of measurement is that independent chimneys are detailed and enumerated while flues are measured in metres with enumerated features as pipelines in Section Y10. Billing this work therefore requires separation of chimneys from flues. In this regard the important word in the SMM is 'independent' as applied to chimneys, and the inference that flues are not independent. A reasonable definition would be that a flue is any portion of a boiler exhaust within or attached to a building structure, but a chimney is any portion which is freestanding either on a foundation or above a roof.

WORKED EXAMPLES

Readers should refer to SMM7 and the Measurement Code while following these examples, which will be cross-referenced where appropriate.

These worked examples are set out in draft bill format which clarifies not only the SMM7 measurement requirements but also the various ancillary matters such as testing and sundries, which may not appear clearly in a basic take-off.

Worked Example 5/1

This example demonstrates the application of the Rules of SMM7 to a typical small-scale heating installation to a single house unit based on the popular micro bore piping installation.

Worked Example 5/2

This example demonstrates the application of the Rules of SMM7 to the new work involved in an extension to the large heating installation of an existing school. (The alterations to existing work concerned in this project are detailed in Chapter 9.) The installation is executed in traditional mild steel screwed piping.

Worked Example 5/3

This example demonstrates the application of the Rules of SMM7 to the special location of a boiler room including the installation of the boiler, hot water storage, steel and copper piping, gas service piping, and the boiler flue and chimney. This bill would be part only of a large project but billed separately in accordance with Rule M2 of Sections Y10/11, Y20–25 and Y50.

Worked Example 5/1: Small Heating Installation in a House

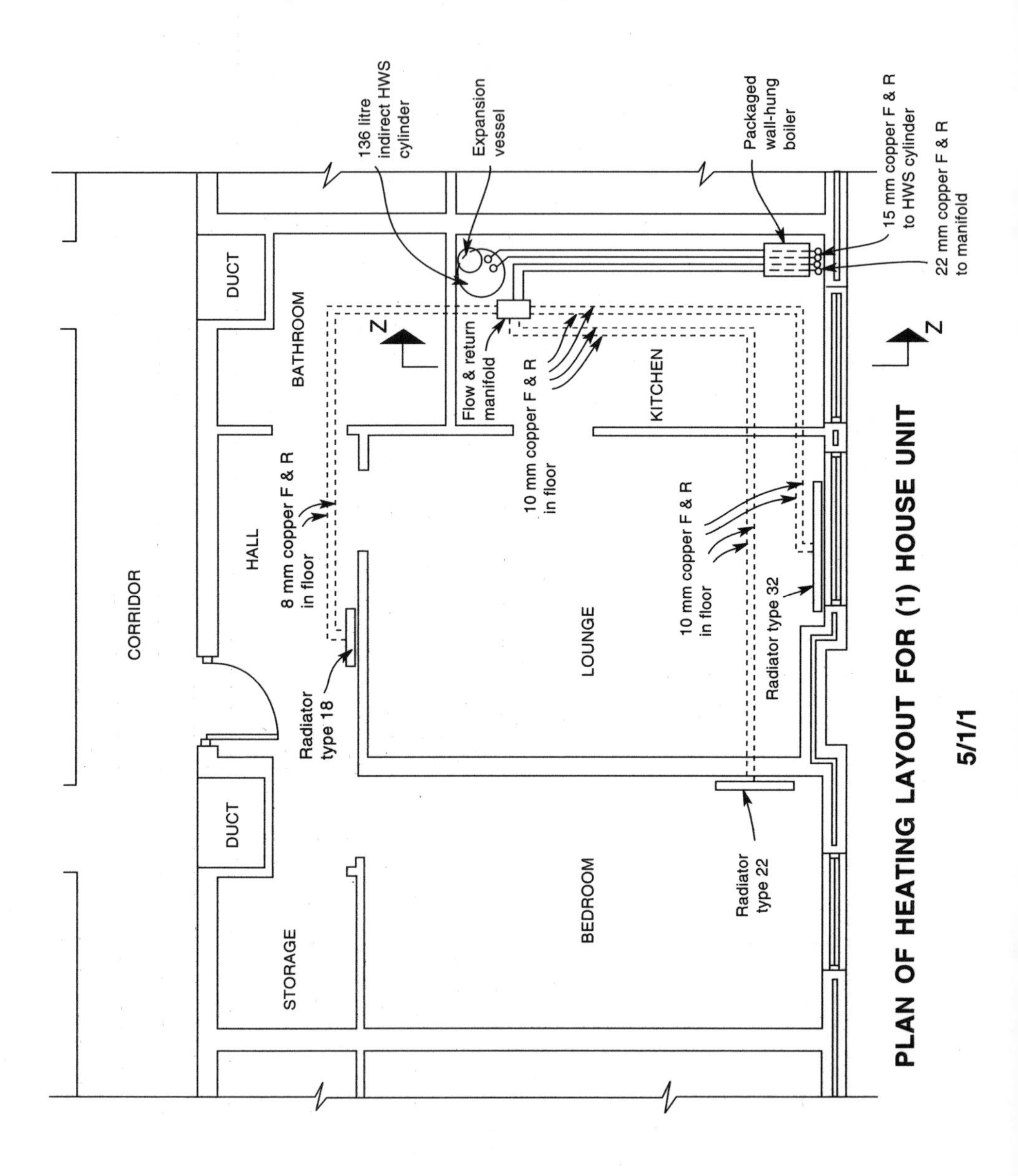

Worked Example 5/1: Small Heating Installation in a House

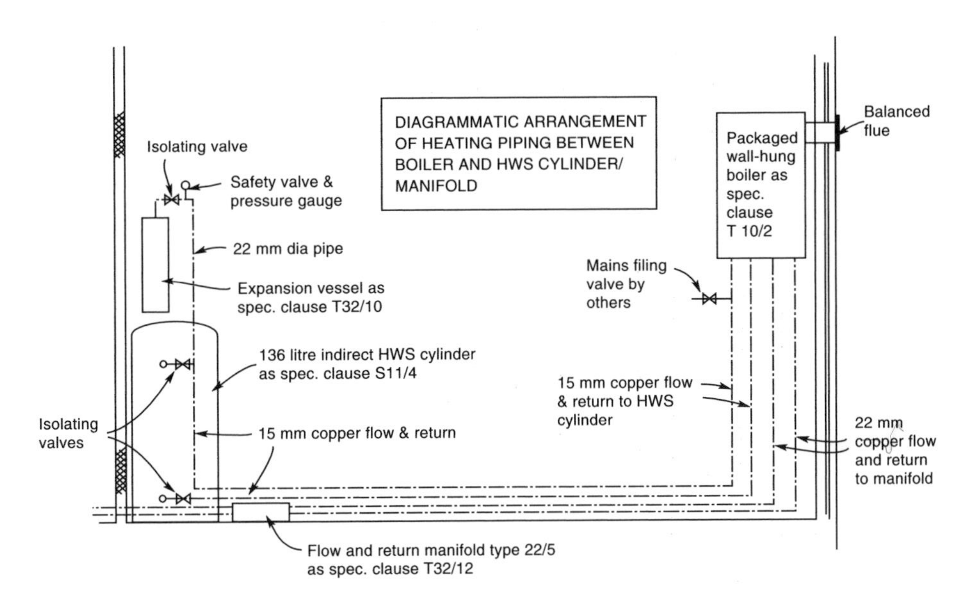

DIAGRAMMATIC SECTION Z–Z

NOTES

Boiler Package unit complete with pumps, 4 × 28 mm male threaded connections for flows & returns.

HWS Cylinder Indirect unit with 2 × 32 mm female threaded primary connections. (Ignore all plumbing connections in this measurement)

Manifold 6 × 10 mm connections both sides, 2 × 22 mm F & R connections.

Walls Generally load bearing concrete blockwork with plaster finish.

Floors Chipboard flooring on 38 × 50 mm battens on sound quilting. All piping to be run between battens below flooring.

Radiators To be RICS enamelled mild steel double panel units as spec. clause T32 /15 of the types noted on plan, fixed with MS brackets plugged and screwed to walls.

Radiator Valves Each radiator to have (1) thermostatic control angle valve and (1) lockshield angle valve.

5/1/2

WORKED EXAMPLE 5/1 – (1) HOUSE UNIT
Abstract Sheets for SMM7 Section T – Mechanical Heating System

ABSTRACT NR 1 – Copper pipe with pipe rings; capillary couplings; masonry background

Location	Pipe		Made Bends		Made Offsets		Fittings 2 ends		Fittings 3 ends		Special Connections					Ancillaries		
											Female bent		Male straight			WGV		Gauge & Safety Valve
	15	22	15	22	15	22	15	22	15	22	28/15	28/22	35/15	22/22	35/22	15	22	
Boiler	2/2.00	2/2.00	2	2	2	2	2	2	1		2	2						
Cylinder	0.75				1		1									1		
Expansion	1.30	1.60		1	1	2	2	3		1			1	1	1		2	1
Totals	6.05	5.60	2	3	4	4	5	5	1	1	2	2	1	1	1	1	2	1

ABSTRACT NR 2 – Copper pipe with capillary couplings; not fixed (laid on floor insulation)

Location	Pipe		Made Offsets		Fittings 2 ends	Ancillaries Manifold
	15	22	15	22	22	
Kitchen floor	2/3.60	2/4.00	2	2	2	1
Totals	7.20	8.00	2	2	2	1

ABSTRACT NR 3 – Copper pipe with capillary couplings; flexible micro bore; not fixed

Location	Pipe		Fittings 2 ends	
	8	10	8	10
Manifold – hall	2/6.00		2	
Manifold – lounge		2/6.25		2
Manifold – bedroom		2/8.50		2
Totals	12.00	29.50	2	4

WORKED EXAMPLE 5/1		SHEET NR 1		
Commentary	Item Nr	Description	Unit	Qty
Project title		*(1) HOUSE UNIT*		
Sequential order of bills		*BILL NR 11*		
Common Arrangement of Work Section heading Heading from Appendix B.		*T – MECHANICAL HEATING SYSTEM* ***T32 Low Temperature Hot Water Heating (Small scale)***		
		Information provided		
This note covers the requirements of Rule P1 of sections Y10–59. Placing this first avoids repetition in each subsection of the bill.		The following represents the installation of a combined small and micro bore heating system to (1) house unit all as detailed on Plans 5/1/1 and 5/1/2.		
Essential heading to indicate rules adopted for measurements following. Table X – normal quality copper tube.		***Y10/11 Pipelines and Pipeline Ancillaries*** *Copper tubing to BS 2871: Part 1 – Table X*		
Two-stage take-off by abstract – see explanation in Chapter 2. Y10.1.1.1.1		Pipes, straight, jointed with copper capillary couplings, fixed to masonry background with brass pipe rings plugged and screwed at centres as specified		
Different diameters in sub-items.	1/1	15 mm diameter	m	6
		Abstract nr 1 6.05		
	1/2	22 mm diameter	m	6
		Abstract nr 1 5.60		
		Pipes, straight, jointed with copper capillary couplings, not fixed		
	1/3	15 mm diameter	m	7
		Abstract nr 2 7.20		
	1/4	22 mm diameter	m	8
		Abstract nr 2 8.00		
Micro bore pipes are laid in place in a similar manner to electric cabling and thus are treated as flexible.		Pipes, flexible, jointed with copper capillary couplings, not fixed		
Y10.1.3.1	1/5	8 mm diameter	m	12
		Abstract nr 3 12.00		
	1/6	10 mm diameter	m	30
		Abstract nr 3 29.50		

WORKED EXAMPLE 5/1			SHEET NR 2	
Commentary	Item Nr	Description	Unit	Qty
Y10.2		*Items extra over copper pipes in which they occur*		
Y10.2.1		Made bends		
Different pipe sizes in sub-items. Items indented once for 'extra over' and once more for sub-items.	2/1	15 mm pipe	nr	2
		Abstract nr 1 2		
	2/2	22 mm pipe	nr	3
		Abstract nr 1 3		
Not in SMM but justified as more labour content than 1 bend but less than 2 bends.		Made offsets		
	2/3	15 mm pipe	nr	6
		Abstract nr 1 4 Abstract nr 2 2		
	2/4	22 mm pipe	nr	6
		Abstract nr 1 4 Abstract nr 2 2		
Y10.2.3.3.2 Simpler approach to measurement of fittings on pipes not exceeding 65 mm diameter.		Fittings, copper capillary pattern, two ends		
	2/5	8 mm pipe	nr	2
		Abstract nr 3 2		
	2/6	10 mm pipe	nr	4
		Abstract nr 3 4		
	2/7	15 mm pipe	nr	5
		Abstract nr 1 5		
	2/8	22 mm pipe	nr	7
		Abstract nr 1 5 Abstract nr 2 2		
Y10.2.3.4.2		Fittings, copper capillary pattern, three ends		
	2/9	15 mm pipe	nr	1
		Abstract nr 1 1		
	2/10	22 mm pipe	nr	1
		Abstract nr 1 1		

WORKED EXAMPLE 5/1		SHEET NR 3		
Commentary	Item Nr	Description	Unit	Qty
Important heading repeated.		*Items extra over copper pipes in which they occur (continued)*		
Y10.2.2.1		Special connection between copper pipe and female threaded equipment with brass compression male straight connection		
	3/1	15 mm pipe to 35 mm equipment	nr	1
		Abstract nr 1 1		
	3/2	22 mm pipe to 22 mm equipment	nr	1
		Abstract nr 1 1		
	3/3	22 mm pipe to 35 mm equipment	nr	1
		Abstract nr 1 1		
		Special connection between copper pipe and male threaded equipment with brass compression female bent connection		
	3/4	15 mm pipe to 28 mm equipment	nr	2
		Abstract nr 1 2		
	3/5	22 mm pipe to 28 mm equipment	nr	2
		Abstract nr 1 2		
Y11.8 + Measurement Code Full value – heading prevents 'extra over' continuing.		*Pipework Ancillaries*		
		Brass wheel gate valves with compression joints to copper pipes		
	3/6	15 mm diameter	nr	1
		Abstract nr 1 1		
	3/7	22 mm diameter	nr	2
		Abstract nr 1 2		
	3/8	Combined safety valve and pressure gauge with compression joints to 22 mm copper pipe, as specification clause T32/51	nr	1
		Abstract nr 1 1		

WORKED EXAMPLE 5/1			SHEET NR 4	
Commentary	Item Nr	Description	Unit	Qty
Y11.8 + Measurement Code		*Pipework Ancillaries (continued)*		
	4/1	Copper flow and return manifold type 22/5 as specification clause T32/12 complete with connections to copper pipe: 2 × 22 mm, 4 × 10 mm, 2 × 8 mm and with 6 × 10 mm blanked off connections	nr	1
		Abstract nr 2 1		
		Y20/25 General Pipeline Equipment		
Y22.1.1.1.1 + 4 + 6	4/2	Packaged gas fired room sealed wall hung boiler as specification clause T10/2 complete with pumps, 4 × 28 mm male threaded connections for flows and returns and balanced flue assembly with outside grille fitted to preformed hole in 300 mm thick outer wall; unit bolted to masonry (gas supply and connection by others)	nr	1
		Kitchen 1		
Y23.1.1.1.4	4/3	Hot water storage cylinder indirect 136 litre copper to BS 699 (Grade 2) as specification clause S11/4 with 2 × 35 mm female threaded primary bosses and 2 × 28 mm female threaded secondary bosses, precoated with 32 mm thick insulation; supported on floor	nr	1
		Kitchen 1		
Y23.1.1.1.4	4/4	Expansion vessel as specification clause T32/10, with 22 mm female threaded connection, unit screwed and plugged to masonry	nr	1
		Kitchen 1		
Y22.1.1.1.1 + 4		Radiators, enamelled mild steel double panel including 20 mm brass air release valves as specification clause T32/15 with mild steel brackets plugged and screwed to masonry		
Different sizes in sub-items.	4/5	Type 18	nr	1
		Hall 1		
	4/6	Type 22	nr	1
		Bedroom 1		
	4/7	Type 32	nr	1
		Lounge 1		

WORKED EXAMPLE 5/1		SHEET NR 5		
Commentary	Item Nr	Description	Unit	Qty
Y22.2.1.1.1 Y22–C4	5/1	Ancillaries for equipment, thermostatic radiator control angle valves 15 mm diameter jointed with demountable couplings to female radiator boss and reduced for 8 or 10 mm copper pipes	nr	3
		All rads 3		
	5/2	Ancillaries for equipment, radiator lockshield angle valves 15 mm diameter jointed as last item	nr	3
		All rads 3		
Assume all 15 & 22 pipes insulated and 8 & 10 pipes insulated under floor only. Y50.1.1.1		***Y50 Thermal Insulation*** <u>*Plastic faced glass fibre sectional insulation minimum 15 mm thick, secured with waterproof adhesive tape*</u>		
	5/3	Pipelines, 8 mm diameter	m	11
		Abstract 3 less 2 × 300 mm above floor 11.40		
	5/4	Pipelines, 10 mm diameter	m	28
		Abstract 3 less 2 × 2 × 300 mm above floor 28.30		
	5/5	Pipelines, 15 mm diameter	m	13
		Abstract nr 1 6.05 Abstract nr 2 7.20		
	5/6	Pipelines, 28 mm diameter	m	14
		Abstract nr 1 5.60 Abstract nr 2 8.00		
Y50.2.1.1	5/7	Extra over pipeline insulation for working around ancillaries – wheel gate valves	nr	3
		Abstract nr 1 – 15 mm 1 22 mm 2		
Y50.2.1.1 – Although manifold is fairly large as a pipe ancillary it is not considered difficult to price as 'extra over' adjoining pipe insulation.	5/8	Extra over pipeline insulation for working around ancillaries – manifold size about 400 × 200 × 100 mm	nr	1
It is not likely that the safety valve would be insulated but if required then measured in similar way to manifold.		Abstract nr 2 1		

WORKED EXAMPLE 5/1			SHEET NR 6	
Commentary	Item Nr	Description	Unit	Qty
		Y51–59 Testing and Sundries		
Y59.1.1	6/1	Marking positions of holes, mortices and chases in the structure for the whole of the foregoing hot water heating system	item	
Y51.4.1	6/2	Testing and commissioning the whole of the foregoing hot water heating system as specification clause Y51/3	item	
P31 Rule M2 P31.20 Cutting or forming holes for services installations are enumerated, see Chapter 2 – 'Builder's Work'. Seems reasonable to take one hole for two very small pipes running parallel.		***P31 Holes, Chases for Hot water heating system***		
	6/3	Cutting or forming holes for pairs of micro bore pipes together not exceeding 55 mm nominal size through concrete blockwork partitions 100 mm thick and making good	nr	3
		F&R to 3 rads 3		
	6/4	Cutting or forming hole for pair of micro bore pipes as last item but through concrete blockwork partition 200 mm thick and making good	nr	1
		F&R to bedroom rad 1		
P31.20.1.1.1	6/5	Cutting or forming hole for balanced flue duct, girth not exceeding 1.00 m rectangular through 300 mm thick cavity walling comprising concrete blockwork inner leaf and facing brick outer leaf and making good	nr	1
		Boiler 1		

Worked Example 5/2: Heating of Laboratory Extension to School

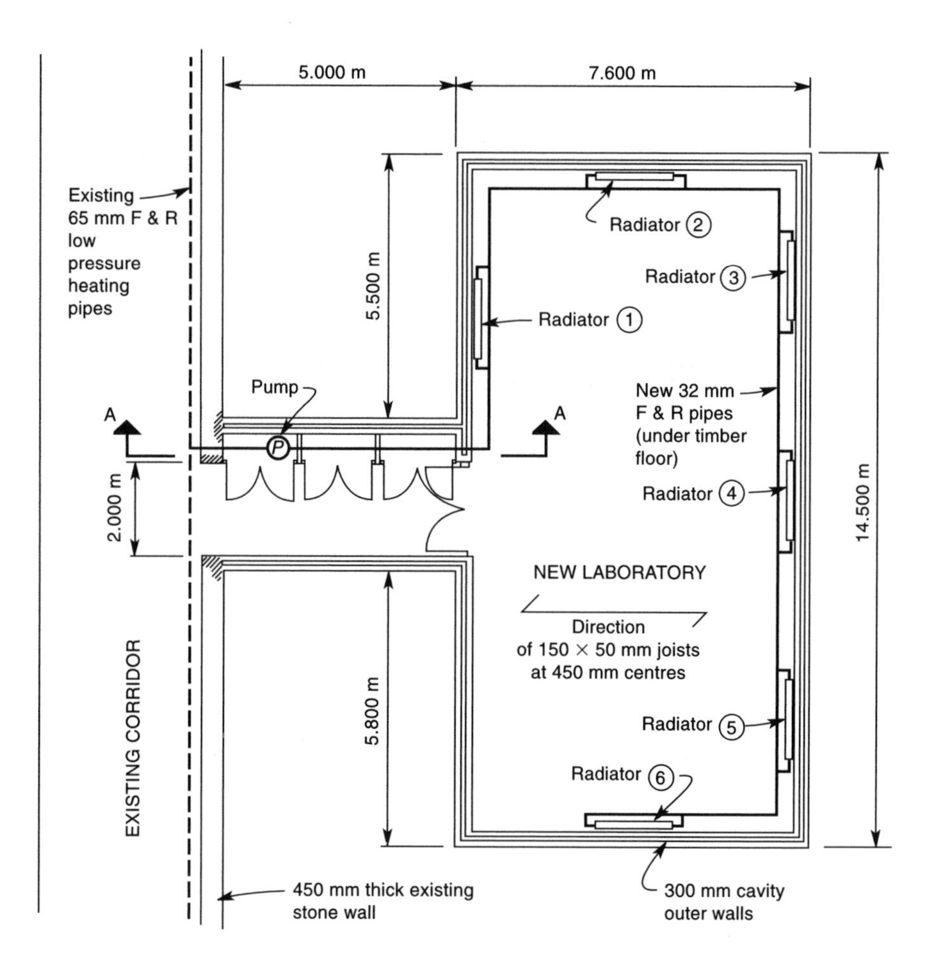

PROPOSED LABORATORY EXTENSION HEATING LAYOUT PLAN

5/2/1

Worked Example 5/2: Heating of Laboratory Extension to School

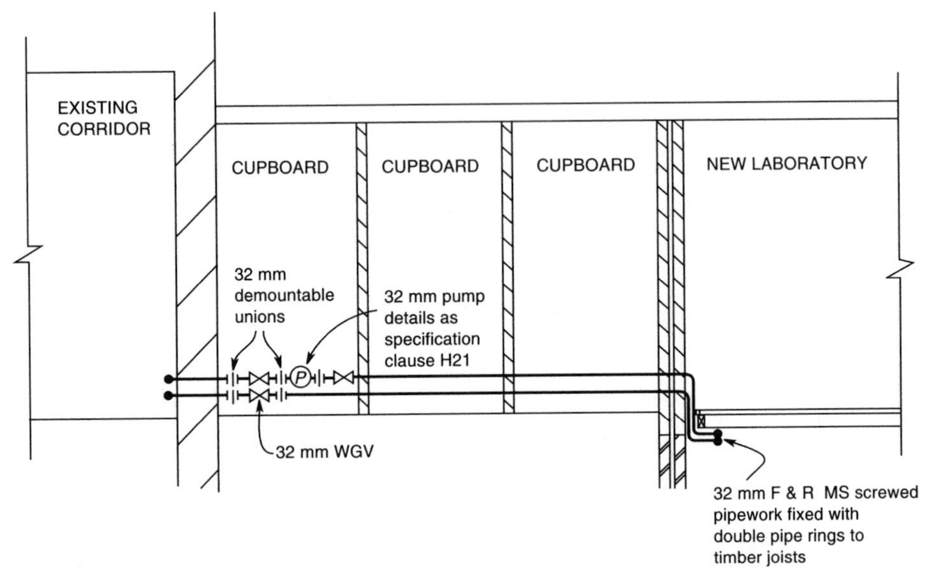

SECTION A–A

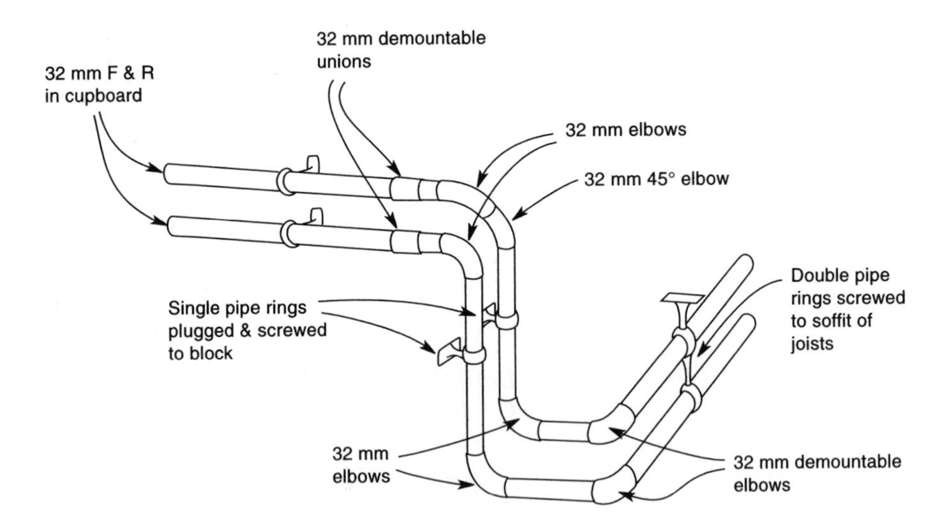

SKETCH OF DROP BELOW FLOOR

5/2/2

Worked Example 5/2: Heating of Laboratory Extension to School

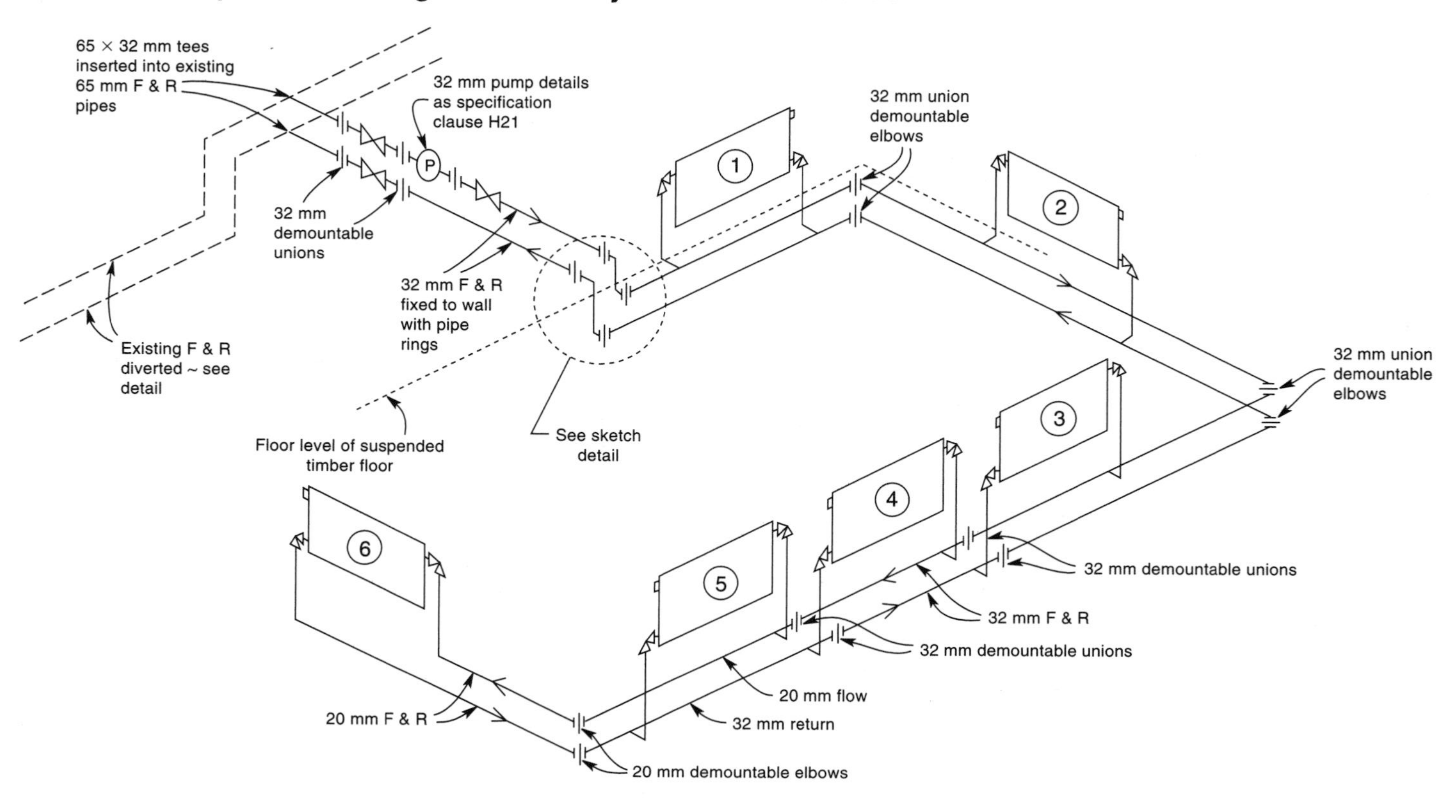

HEATING LAYOUT – DIAGRAMMATIC LAYOUT

5/2/3

Worked Example 5/2: Heating of Laboratory Extension to School

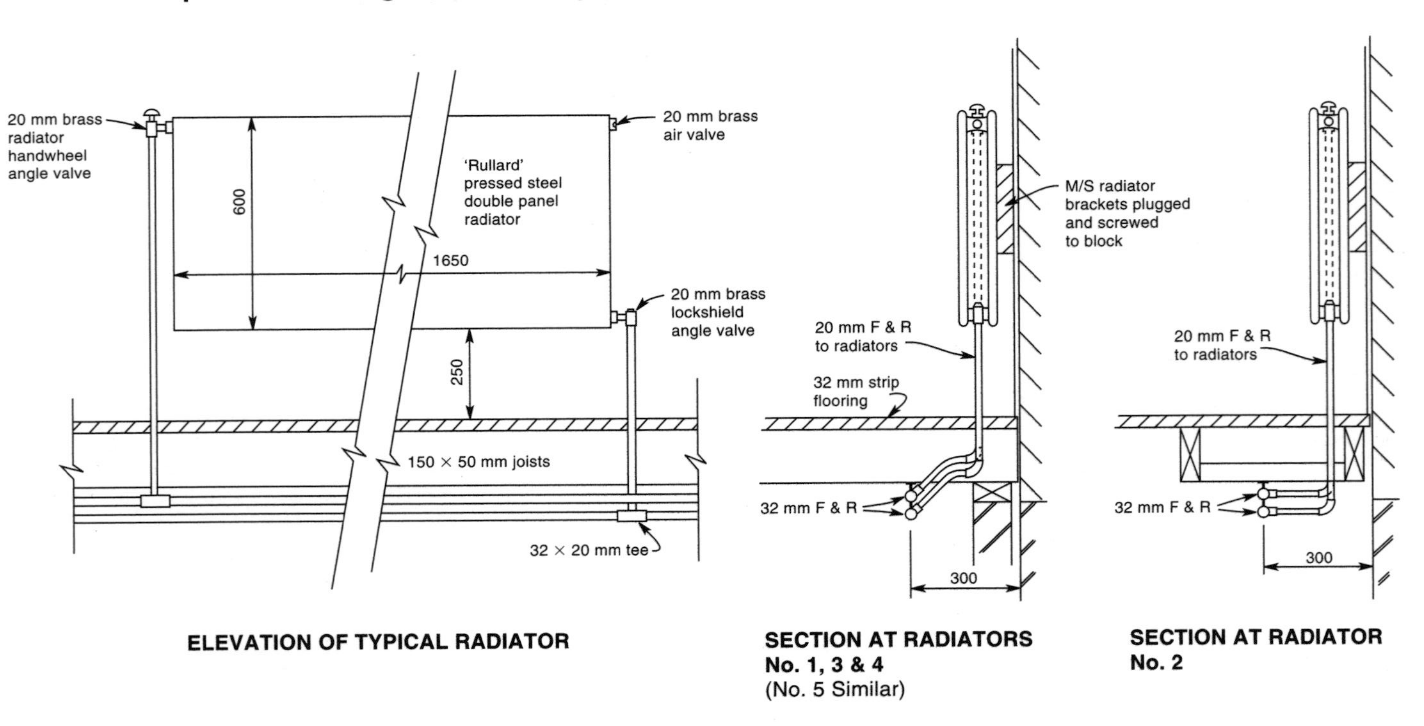

5/2/4

Worked Example 5/2: Heating of Laboratory Extension to School

NEW CONSTRUCTION

Outer walls: 300 mm thick cavity construction comprising rendered brick outer skin and insulating blockwork inner skin.

Partitions: 100 mm thick insulating blockwork.

Plaster finish: 2 coat plasterwork to all walls except in cupboards which are pointed blockwork finish.

Floors: New Link Corridor – solid concrete sandwich construction with cement and sand screed.
New Laboratory – suspended timber construction with 150 × 50 mm softwood joists at 450 mm centres and 32 mm T & G hardwood strip flooring.

SPECIFICATION NOTES

Existing system: existing system comprises gas fired low pressure heated water heating installation.

New pipework: to be mild steel screwed pipework to BS 1387: 1985 medium grade.

Screwed joints to be sealed with PTFE. Pipe fittings to be mild steel screwed to BS 143 & 1256:1986.

All pipework under timber floors and within ducts to be insulated with sectional glass fibre insulation secured with self-adhesive tape.

Radiators: to be 'Rullard' standard pressed steel double panel radiators capable of being readily demounted for maintenance.

Pump: water circulating pump to conform to specification clause H21 with female ends (electrical connections elsewhere included).

5/2/5

WORKED EXAMPLE 5/2 – LABORATORY EXTENSION
Abstract Sheets for SMM7 Section T – Mechanical Heating System

ABSTRACT NR 1 – 32 mm MS pipe with pipe rings; screwed couplings; masonry background

Location	Pipe	Fittings 2 ends	Special Connections		Ancillaries WGV		Equipment Pump
			DU	DE	1DU	2DU	
Cupbs F+R Drop F+R Deduct – F valves+pump R valve	2/5.35 1.40 Dt0.50 Dt0.10	5	2	2	1	2	1
Totals	11.50	5	2	2	1	2	1

ABSTRACT NR 2 – MS pipe fixings measured elsewhere; screwed couplings

Location	Pipe		Fittings 3 ends 32	Special Connections		
	20	32		DU 32	DE 20	DE 32
Under timber floor						
West F+R		2/5.80	2			2
North F+R		2/6.40	2			
East F	2.50	10.80	3	2		2
R	0.80	12.50	3	2	2	
South F+R	2/3.60					
Totals	10.50	47.70	10	4	2	4

Abbreviations: DU = Demountable Union
DE = Demountable Elbow

WORKED EXAMPLE 5/2 – LABORATORY EXTENSION (cont.)

ABSTRACT NR 3 – 20 mm MS pipe with pipe rings; screwed couplings; timber background

Location	Pipe
Return rad 6	1.70
Totals	1.70

Commentary: Although this category does not merit an abstract, it is included for uniformity and to prevent it being omitted by error.

ABSTRACT NR 4 – 20 mm MS pipe not fixed; screwed couplings (short lengths to rads)

Location		Pipe	Made Bends	Made Offsets	Fittings 2 ends
Rads 1, 3, 4, 5	flows	4/1.20	4		4
	returns	4/0.60	4		4
Rad 2	flow	1.30		1	1
	return	0.75		1	1
Rad 6	flow	1.05		1	1
	return	0.45		1	1
Totals		10.75	8	4	12

WORKED EXAMPLE 5/2 SHEET NR 1

Commentary	Item Nr	Description	Unit	Qty
Project title		*LABORATORY EXTENSION*		
Sequential order of bills		*BILL NR 10*		
Common Arrangement of Work Section heading		*T – MECHANICAL HEATING SYSTEM*		
Heading from Appendix B.		***T31 Low Temperature Hot Water Heating***		
		Information provided		
This note covers the requirements of Rule P1 of sections Y10–59. Placing this first avoids repetition in each subsection of the bill. Clarification that this bill is solely concerned with new work as work within existing buildings must be so defined – SMM7 – General Rules 7 & 13. (Alterations in Chapter 9 – Worked Example 9/2.)		The following represents the installation of a heating system within the new laboratory extension to an existing school as detailed on Plans 5/2/1–5. *Note:* all alterations to pipework etc. within the existing building in connection with this installation are measured elsewhere.		
Essential heading to indicate rules adopted for measurements following.		***Y10/11 Pipelines and Pipeline Ancillaries***		
		Mild steel screwed pipework to BS 1387, table 4, medium weight, jointed with plain sockets with PTFE compound		
Two stage take-off by abstract – see explanation in Chapter 2. Y10.1.1.1.1	1/1	Pipes, straight, fixed to timber background with MS pipe rings at centres as specified; 20 mm diameter	m	2
		Abstract nr 3 1.70		
	1/2	Pipes, straight, fixed to masonry background with MS pipe rings plugged and screwed at centres as specified; 32 mm diameter	m	12
		Abstract nr 1 11.50		
Y10 Rule M6 – more than 1 pipe supported on fabricated support then measured as P31.30.		Pipes, straight, fixings measured elsewhere		
Different diameters in sub-items.	1/3	20 mm diameter	m	11
		Abstract nr 2 10.50		
	1/4	32 mm diameter	m	48
		Abstract nr 2 47.70		
MS pipes are mechanically strong and in such location are rarely fixed thus item makes this clear.	1/5	Pipes, straight, not fixed (short lengths to radiators); 20 mm diameter	m	11
		Abstract nr 4 10.75		

WORKED EXAMPLE 5/2		SHEET NR 2		
Commentary	**Item Nr**	**Description**	**Unit**	**Qty**
Y10.2		*Items extra over MS pipes in which they occur*		
Y10.2.1		Made bends		
Different pipe sizes in sub-items. Items indented once for 'extra over' and once more for sub-items.	2/1	20 mm pipe	nr	8
		Abstract nr 4 8		
No 32 mm bends shown on plans but may occur in reality owing to possible obstructions at installation – often taken to have rate in bill.	2/2	32 mm pipe	nr	2
		(Say) 2		
Not in SMM but justified as more labour content than 1 bend but less than 2 bends.	2/3	Made offsets; 20 mm pipe	nr	4
		Abstract nr 4 4		
Y10.2.3.3.2 Simpler approach to measurement of fittings on pipes not exceeding 65 mm diameter.		Fittings, MS screwed pattern, two ends		
	2/4	20 mm pipe	nr	12
		Abstract nr 4 12		
	2/5	32 mm pipe	nr	5
		Abstract nr 1 5		
Y10.2.3.4.2	2/6	Fittings, MS screwed pattern, three ends; 32 mm pipe	nr	10
		Abstract nr 2 10		
Y10.2.2.1 – Rule D2 – joints which differ from those in running length.	2/7	Special joints, MS screwed demountable unions; 32 mm pipe	nr	6
		Abstract nr 1 2 Abstract nr 2 4		
		Special joints, MS screwed demountable elbows		
	2/8	20 mm pipe	nr	2
		Abstract nr 2 2		
	2/9	32 mm pipe	nr	6
		Abstract nr 1 2 Abstract nr 2 4		
Broken line to end 'extra over'.		- - - - - - - - - - - - - -		

WORKED EXAMPLE 5/2			SHEET NR 3	
Commentary	Item Nr	Description	Unit	Qty
Y11.8 + Measurement Code Full value. Method of jointing stated.		*Pipework ancillaries*		
	3/1	MS wheel gate valve once screw jointed and once jointed with MS demountable union to MS screwed pipe; 32 mm diameter	nr	1
		Abstract nr 1 1		
	3/2	MS wheel gate valve twice jointed with MS demountable unions to MS screwed pipe; 32 mm diameter	nr	2
		Abstract nr 1 2		
Y11.11.1.1.2	3/3	Pipe sleeves through walls, length not exceeding 300 mm in MS for 32 mm pipe, handed to builder for fixing	nr	6
		Cupb. partitions 2 × 2 Cupb./Lab. 300 cav. 2		
Y11.12.*.1 Different sizes in sub-items.		Floor plates, split MS pattern, screwed to timber		
	3/4	20 mm pipe	nr	12
		Rads F&R 6 × 2		
	3/5	32 mm pipe	nr	2
		F&R drop Lab. 2		
		Y20/25 General Pipeline Equipment		
Y20.1.1.1.1	3/6	Water circulating pump for 32 mm pipe as specification clause H21 with female ends (electrical connections measured elsewhere)	nr	1
		Abstract nr 1 1		
Y22.1.1.1.1 + 6	3/7	'Rullard' pressed steel double panel radiators, each with 1 × 20 mm brass aircock, fixed with concealed MS brackets plugged and screwed to masonry background; 2.96 m² heating surface, 1650 × 600 mm	nr	6
		Lab. rads 1–6 6		
Y22.2.1.1	3/8	Ancillaries for equipment, 20 mm brass radiator control wheel angle valves, once jointed to MS screwed pipe and once jointed to radiator with brass disconnecting union	nr	6
		All rads 6		

WORKED EXAMPLE 5/2		SHEET NR 4		
Commentary	Item Nr	Description	Unit	Qty
		Y20/25 General Pipeline Equipment (cont.)		
Y22.2.1.1	4/1	Ancillaries for equipment, 20 mm brass radiator lockshield angle valves, jointed as last item	nr	6
		All rads 6		
Y22.10.1	4/2	Disconnecting, setting aside and refixing (6) pressed steel radiators for convenience of Painting Contractor	item	
Specification states all pipes Under Floor to be insulated. Y50.1.1.1		***Y50 Thermal Insulation***		
		Plastic faced glass fibre sectional insulation minimum 15 mm thick, secured with waterproof self-adhesive tape		
	4/3	Pipelines, 20 mm diameter	m	16
		Abstract nr 2 10.50 Abstract nr 3 1.70 Abstract nr 4 only F&R under floor 2/6/0.35		
	4/4	Pipelines, 32 mm diameter	m	48
Abstract nr 1 nil insulation – all above floor.		Abstract nr 2 47.70		
		Y51–59 Testing and Sundries		
Y59.1.1	4/5	Marking positions of holes, mortices and chases in the structure for the whole of the foregoing hot water heating system	item	
Y51.4.1	4/6	Testing and commissioning the whole of the foregoing hot water heating system as per specification clause Y51/4	item	
P31 Rule M2		***P31 Holes, Chases, Supports for Hot water heating system***		
P32.20 Cutting or forming holes for services installations are enumerated, see Chapter 2 – 'Builder's Work'. Pipe sleeve supply in item 3/3.	4/7	Cutting or forming holes for pipes not exceeding 55 mm nominal size through concrete blockwork partitions 100 mm thick, take delivery of and build in pipe sleeve and making good	nr	4
		Cupb. partitions 4		
	4/8	Cutting or forming hole for pipes as last item but through concrete blockwork cavity wall 300 mm thick, take delivery of and build in pipe sleeve and making good	nr	2
		Cupb./Lab. 2		

WORKED EXAMPLE 5/2			SHEET NR 5	
Commentary	Item Nr	Description	Unit	Qty
P31.30.3.1 – Supports for more than one pipe separately measured from items for pipes.		Soffit pipe supports comprising pipe rings and backplates screwed to timber background		
Sub-items for size variants.	5/1	Supports for 2 × 20 mm pipes	nr	3
		South-east Lab. 3		
	5/2	Supports for 1 × 20 mm and 1 × 32 mm pipe	nr	1
		East Lab. 1		
	5/3	Supports for 2 × 32 mm pipes	nr	16
		West Lab. 4 North Lab. 4 East Lab. 8		

Worked Example 5/3: Boiler House Layout

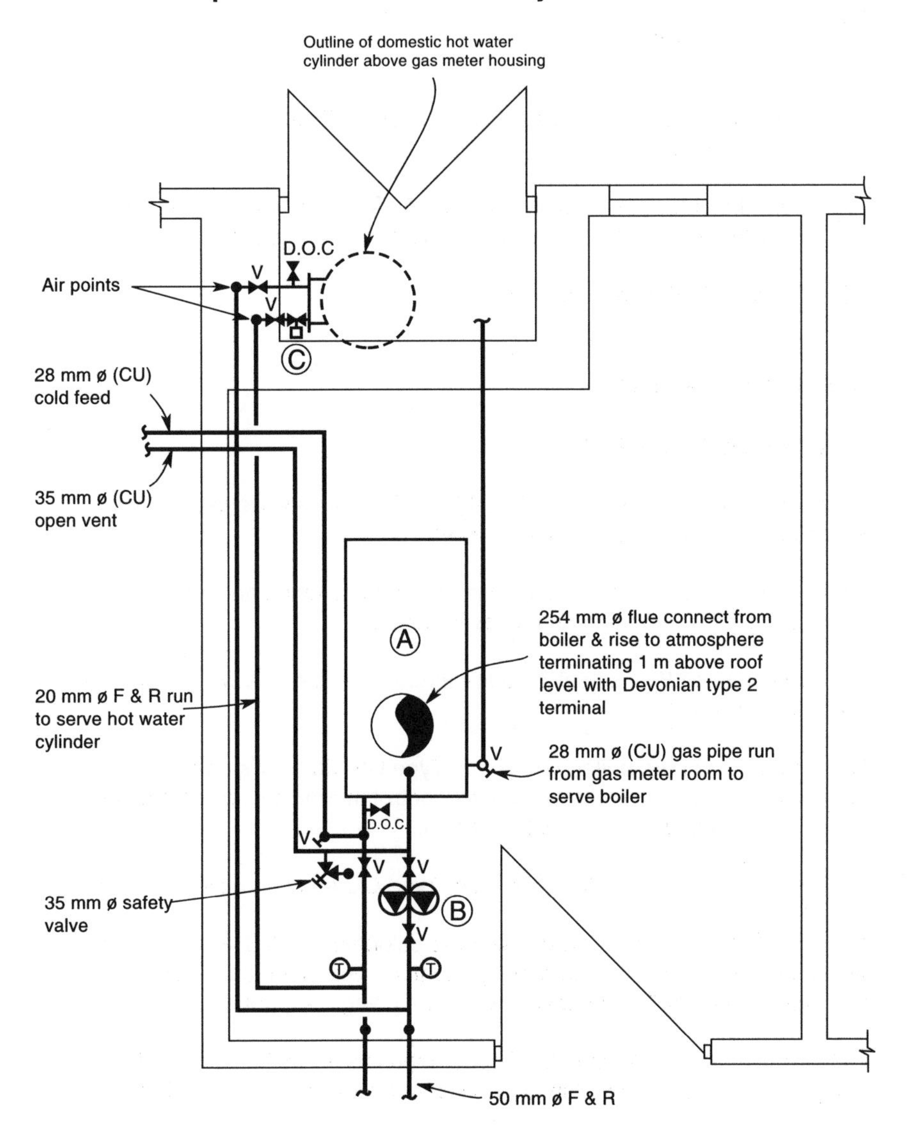

BOILERHOUSE LAYOUT

Legend:

F & R – Flow & Return
V – Isolating Valve
D.O.C. – Drain Cock
Ⓣ – Temperature gauge

5/3/1

Worked Example 5/3: Boiler House Layout

PLANT SCHEDULE

(A) Hamworthy Ltd. UR Series, Gas fired boiler Type UR430, output 95.8 kW.

(B) Grundfos Ltd. Twin head pump set, Type UPCD 40 – 120, Duty 2 l/s at 60 kPa suitable for 240 V/1 ph/50 Hz electrical supply.

(C) 20 mm Dia. 2 port motorised valve.

SPECIFICATION NOTES

HEATING INSTALLATION

Pipework to be Steel pipes to BS 1387, medium quality, weldable class, weldable Class B fittings, in pipe brackets.

Drain valves to be Gunmetal drain valve with hose union and key valve, Ref.81HU with wheel head and screwed ends.

Gate valves to be Bronze solid wedge gate valve, Ref.33X with wheel head and screwed ends.

Boiler to be Gas fired boiler, UR Series Type UR430 floor standing conventional flue boiler with boiler limit/control thermostat, automatic gas valve, gas isolating valve, 95.8 kW output.

Flue to be 254 mm diameter boiler flue, Type QC and connecting to boiler.

Calorifier to be Copper hot water storage calorifier to BS 853, Part II, supplied with 25 mm safety valve with drip pipe to floor, Gunmetal drain cock with hose union, 227 litre capacity with metal angle supports.

Cold Feed and Open Vent to be Copper pipes to BS 2871, Part I, Table X, Copper non-dezincifiable fittings to BS 864, Part 2, compression Type B joints.

5/3/2

WORKED EXAMPLE 5/3			SHEET NR 1	
Commentary	Item Nr	Description	Unit	Qty
Project title		*NEW AIRPORT BUILDING*		
Sequential order of bills		*BILL NR 12*		
Common Arrangement of Work Section heading		*T – MECHANICAL HEATING SYSTEM*		
Heading from Appendix B.		***T31 Low Temperature Hot Water Heating***		
		Information provided		
This note covers the requirements of Rule P1 of sections Y10–59. Placing this first avoids repetition in each subsection of the bill.		The following represents the installation of a heating system within the new Airport Arrivals Building as detailed on Plans 5/3/1 and 5/3/2.		
Assume measurement of all non-plant-room work would be measured in the usual way for this project. All work in the boiler house is billed separately in this example:		*Y10/11 Pipelines and Pipeline Ancillaries* *Y20/25 General Pipeline Equipment* *Y50 Thermal Insulation* *Y51–59 Testing and Sundries* *P31 Holes, Chases, Supports for Hot water heating system*		
Y10/11 & Y20/25 Rule M2 special location requirement + optional specific location in brackets.		**Work in Plant Room (Boiler House)**		
Essential heading to indicate rules adopted for measurements following.		***Y10/11 Pipelines and Pipeline Ancillaries***		
		Mild steel screwed pipework to BS 1387, table 4, medium weight, jointed with plain sockets with PTFE compound		
Different diameters in sub-items.		Pipes, straight, fixed to masonry background with MS pipe rings plugged and screwed at centres as specified		
Direct take off as straightforward run of pipes within boiler room.	1/1	20 mm diameter	m	6
		F&R dom. hot horiz. 2/2.20 vert. 2/1.00		
	1/2	50 mm diameter	m	4
		F&R vert. 2/1.80		
		Pipes, straight, fixed to metal background with MS pipe rings at centres as specified		
	1/3	20 mm diameter	m	1
		Near boiler – flow 0.70 return 0.50		
	1/4	50 mm diameter	m	3
		Near boiler – flow 1.10 return 1.95		

WORKED EXAMPLE 5/3			SHEET NR 2	
Commentary	Item Nr	Description	Unit	Qty
Pipe items (cont.)		*Mild steel screwed pipework (cont.)*		
MS pipes are mechanically strong and in many locations are not fixed and item makes this clear.	2/1	Pipes, straight, not fixed (short lengths to equipment); 20 mm diameter	m	1
		At calorifier F&R 1.40		
Y10.2		*Items extra over MS pipes in which they occur*		
Y10.2.1	2/2	Made bends; 20 mm pipe	nr	4
		(say) 4		
Not in SMM but justified as more labour content than 1 bend but less than 2 bends.	2/3	Made offsets; 20 mm pipe	nr	4
		(say) 4		
Y10.2.3.3.2 Simpler approach to measurement of fittings on pipes not exceeding 65 mm diameter.		Fittings, MS screwed pattern, two ends		
	2/4	20 mm pipe	nr	4
		F&R 4		
	2/5	50 mm pipe	nr	2
		F&R 2		
Y10.2.3.4.2		Fittings, MS screwed pattern, three ends		
	2/6	20 mm pipe	nr	3
		Tees for ancillaries 3		
	2/7	50 mm pipe	nr	7
		Feed & expansion + calorifier 4 Tees for ancillaries 3		
Y10.2.2.1 – Rule D2 – joints which differ from those in running length.	2/8	Special joints, MS screwed demountable unions; 50 mm pipe	nr	2
		(say) 2		
	2/9	Special joints, MS screwed demountable elbows		
	2/10	20 mm pipe	nr	4
		F&R 4		
	2/11	50 mm pipe	nr	4
		F&R 4		

WORKED EXAMPLE 5/3			SHEET NR 3	
Commentary	**Item Nr**	**Description**	**Unit**	**Qty**
Important heading repeated.		*Item extra over MS pipes in which they occur (cont.)*		
Y10.2.2.1		Special connections between pipe and female threaded equipment with MS male demountable straight connection		
	3/1	20 mm pipe	nr	2
		Calorifier F&R 2		
	3/2	50 mm pipe	nr	2
		Boiler F&R 2		
Y11.8 + Measurement Code		*Pipework Ancillaries*		
Full value. Method of jointing stated.	3/3	MS wheel gate valves once screw jointed and once jointed with MS demountable union to MS screwed pipe; 20 mm diameter	nr	2
		Calorifier 2		
	3/4	MS wheel gate valve twice jointed with MS demountable unions to MS screwed pipe; 50 mm diameter	nr	3
		Flow at pump 2 Return 1		
Y11.8.1.1.3	3/5	Motorised valve (two port) as specification clause T11/24 twice jointed with MS demountable unions to MS screwed pipe; 20 mm diameter (electrical connections measured elsewhere)	nr	1
		Calorifier 1		
Y11.8.1.1	3/6	Air release points comprising 20 mm diameter gunmetal air valve with key, screw jointed to previously measured MS fittings	nr	2
		Calorifier 2		
	3/7	Gunmetal drain down cocks 20 mm diameter with hose union and key valve, screw jointed to already measured MS fittings	nr	2
Rule Y10/M5 – 50 × 50 × 20 tee taken.		To 20 mm pipe 1 To 50 mm pipe 1		
	3/8	Temperature gauges as specification clause Y11/26, screw jointed to already measured MS fittings	nr	2
		To 50 mm F&R 2		

WORKED EXAMPLE 5/3			SHEET NR 4	
Commentary	Item Nr	Description	Unit	Qty
		Y20/25 General Pipeline Equipment		
Y20.1.1.1.1	4/1	Water circulating pump twin head pattern for 50 mm pipe as specification clause Y20/23 set on anti-vibration pad and bolted to metal (electrical connections measured elsewhere)	nr	1
		Flow 1		
Y23.1.1.1.1 + 2	4/2	Copper hot water storage calorifier 227 litre as specification clause Y20/33 complete with 25 mm safety valve, 25 mm drain down cock, set on metal supports supplied by others	nr	1
		Above gas meter house 1		
Y22.1.1.1.1 + 2	4/3	Gas fired boiler 95.8 kW output floor mounted, conventional flue as specification clause Y20/35 complete with boiler stat, gas isolating valve and automatic gas valve	nr	1
		'A' on plan 1		
Y20.6.1 Pump sited about 1 m above boiler room floor. Assume drawing provided with tender documents.	4/4	Support for pump comprising fabricated MS angle frame overall size 850 × 450 × 975 mm as component drawing nr Y20/15	nr	1
		Pump 1		
Y22.7 Freestanding section of boiler chimney – refer to explanatory paragraph in text of this chapter. Y22.7.1.0.12	4/5	Independent vertical stainless steel double skin chimney, jointed etc. as specification clause Y20/50; 254 mm internal diameter and total 1000 mm high above roof including combined flat roof seal and flashing skirt and Devonian Type 2 terminal; jointed to flue measured below	nr	1
		Roof 1		
Y22.7 + Rule M3 – Flues as Y10/11. Refer to explanatory paragraph in text of this chapter.		*Y10/11 Stainless steel double skin boiler flue jointed etc. as specification for chimney above as clause Y20/50*		
Y10.1.1.1 (self-supporting to roof)	4/6	Pipes, straight, 254 mm internal diameter not fixed	m	3
Includes length of double bend above boiler.		Boiler – roof 2.80		
Y10.2		Extra over 254 mm pipe		
Y10.2.4.6 – fittings over 65 mm diam. given in detail.	4/7	Obtuse bends	nr	2
		Double bend 2		
Y10.2.2.1	4/8	Special connection to boiler spigot with sealing clamp	nr	1
		1		

WORKED EXAMPLE 5/3			SHEET NR 5	
Commentary	Item Nr	Description	Unit	Qty
Although the cold feed and open vent pipes are very similar to the piped supply installation they should be taken in section T as part of heating. Direct take off as straightforward run of pipes within boiler room. Y10.1.1.1.1		*Y10/11 Copper feed and expansion pipes to heating system – copper tubing to BS 2871: Part 1 – Table X* Pipes, straight, jointed with brass compression couplings, fixed to masonry background with brass pipe rings plugged and screwed at centres as specified		
Different diameters in sub-items.	5/1	28 mm diameter	m	2
		Feed vertical 1.85		
	5/2	35 mm diameter	m	2
		Vent vertical 2.00		
Y10 Rule M6 – more than 1 pipe supported on fabricated support then measured as P31.30.		Pipes, straight, jointed with brass compression couplings, fixings measured elsewhere		
	5/3	28 mm diameter	m	2
		Feed horizontal 2.20		
	5/4	35 mm diameter	m	2
		Vent horizontal 2.30		
Y10.2		*Items extra over copper pipes in which they occur*		
Y10.2.1		Made bends		
Different pipe sizes in sub-items. Items indented once for 'extra over' and once more for sub-items.	5/5	28 mm pipe	nr	2
		(say) 2		
	5/6	35 mm pipe	nr	2
		(say) 2		
Not in SMM but justified as more labour content than 1 bend but less than 2 bends.		Made offsets		
	5/7	28 mm pipe	nr	1
		(say) 1		
	5/8	35 mm pipe	nr	1
		(say) 1		

WORKED EXAMPLE 5/3			SHEET NR 6	
Commentary	Item Nr	Description	Unit	Qty
Important heading repeated.		*Items extra over copper pipes in which they occur (cont.)*		
Y10.2.3.3.2 Simpler approach to measurement of fittings on pipes not exceeding 65 mm diameter.		Fittings, brass compression pattern, two ends		
	6/1	28 mm pipe	nr	4
		Bent couplings 4		
	6/2	35 mm pipe	nr	4
		Bent couplings 4		
	6/3	Fitting, brass compression pattern, three ends; 28 mm pipe	nr	1
		Tee for safety valve 1		
Y10.2.2.1		Special connection between copper pipe and female threaded MS fitting with brass compression male straight connection		
50 × 50 × 32 tee taken with MS pipe.	6/4	28 mm copper pipe to 32 mm MS socket	nr	1
		Feed to return 1		
50 × 50 × 40 tee taken with MS pipe.	6/5	35 mm copper pipe to 40 mm MS socket	nr	1
		Vent to flow 1		
Y11.8 + Measurement Code Full value – heading prevents 'extra over' continuing.		*Pipework Ancillaries*		
	6/6	Brass wheel gate valve with compression joints to copper pipes; 28 mm diameter	nr	1
		Feed 1		
35 × 35 × 35 tee taken above and the minimal length of pipe between them can be ignored.	6/7	Safety valve as specification clause Y11/37 35 mm diameter with compression joints to 35 mm pipe	nr	1
		Vent 1		
As the gas supply in this project only serves boiler it is appropriate to include here but if serving several appliances preferably taken as separate section.		*S32 Natural Gas Installation – serving boiler Y10/11 Copper tubing to BS 2871: Part 1 – Table X*		
Direct take off as straightforward run of pipes within boiler room. Y10.1.1.1.1	6/8	Pipes, straight, jointed with brass compression couplings, fixed to masonry background with brass pipe rings plugged and screwed at centres as specified; 25 mm diameter	m	3
		Meter – boiler 2.60		

WORKED EXAMPLE 5/3		SHEET NR 7		
Commentary	Item Nr	Description	Unit	Qty
Y10.2		*Items extra over copper pipes in which they occur*		
Y10.2.3.3.2	7/1	Fittings, brass compression pattern, two ends; 28 mm pipe	nr	4
		Bent couplings 4		
Y10.2.2.1.1 Assume connection at meter executed by others.	7/2	Special connection between 28 mm copper pipe and 25 mm MS male gas appliance with brass compression female bent connection	nr	1
		Boiler connection 1		
Y11.8 + Measurement Code Full value – heading prevents 'extra over' continuing.		*Pipework Ancillaries*		
	7/3	Brass gas lever valve with compression joints to copper pipes; 28 mm diameter	nr	1
		Near boiler 1		
Y11.11.1.1.2	7/4	Pipe sleeve through wall, length not exceeding 300 mm in copper for 28 mm pipe, handed to builder for fixing	nr	1
		Boiler room/gas meter 1		
Assume only pipes insulated – pipe ancillaries not insulated. Y50.1.1.1		***Y50 Thermal Insulation***		
		Plastic faced glass fibre sectional insulation minimum 25 mm thick, secured with waterproof self-adhesive tape		
	7/5	Pipelines, 20 mm diameter	m	9
Sizes from pipe items.		F&R dom. hot 6.40 1.20 1.40		
Vent insulated (feed not insulated).	7/6	Pipelines, 35 mm diameter	m	4
		Vent 2.00 2.30		
	7/7	Pipelines, 50 mm diameter	m	7
		F&R 3.60 3.05		
Proprietary insulating jacket.		*Preformed glass fibre quilt insulation with plastic face as specification clause Y50/61 manufactured by Insulpak plc*		
Y50.1.4.2	7/8	Jacket to suit 227 litre calorifier as described in item 4/2; quilt minimum 50 mm thick with metal strap fixings	nr	1
		1		

WORKED EXAMPLE 5/3			SHEET NR 8	
Commentary	Item Nr	Description	Unit	Qty
		Y51–59 Testing and Sundries		
Y59.1.1 Location included as Y51–59 will also be measured for non-plant-room work. (Optionally these items could be combined provided clearly described.)	8/1	Marking positions of holes, mortices and chases in the structure for the whole of the foregoing hot water heating system within plant room (Boiler House) including feed and vent pipes	item	
Y59.1.1 – Kept separate as different type of work	8/2	Marking positions of holes, mortices and chases in the structure for the whole of the foregoing natural gas installation within plant room (Boiler House and Gas Meter Housing)	item	
Y51.4.1	8/3	Testing and commissioning the whole of the foregoing hot water heating system within plant room (Boiler House) as specification clause Y51/4	item	
Y51.4.1 – Kept separate as different testing criteria from heating.	8/4	Testing and commissioning the whole of the foregoing natural gas installation within plant room (Boiler House and Gas Meter Housing) as specification clause Y51/5	item	
P31 Rule M2 P31.20 Cutting or forming holes for services installations are enumerated, see Chapter 2 – 'Builder's Work'.		***P31 Holes, Chases, Supports for Hot water heating system and natural gas installation***		
Pipe sleeve supply in item 7/4. Pipe holes to other rooms taken with non-plant-room work.	8/5	Cutting or forming hole for pipe not exceeding 55 mm nominal size through concrete blockwork wall 200 mm thick, take delivery of and build in pipe sleeve and making good Boiler house/gas meter 1	nr	1
Roof deck *in situ* before flue – built up roofing executed after, see Chapter 2 'Builder's Work'.	8/6	Cutting hole for boiler flue exceeding 110 mm nominal size through metal roof deck (roof felt flashing measured elsewhere) 1	nr	1
P31.30.3.1 – Supports for more than one pipe separately measured from item for the pipe.	8/7	Soffit pipe supports comprising pipe rings and backplates screwed to timber background; for 1 × 28 mm and 1 × 35 mm pipes Feed & Vent 3	nr	3

6 Measurement of Ventilation/Air Conditioning Systems

Introduction

The installations covered by this classification can vary from the simplest air supply/extract system to full-scale air conditioning of the most sophisticated nature. These systems encompass the handling of untreated air, heated and/or cooled air, treated air and hybrid systems.

The components of all such air handling installations are basically similar, irrespective of the level of sophistication of the design; thus the approach adopted to measurement can be the same for all types of system.

In Chapter 2 the general approach to billing and measurement of building services was covered, indicating three basic components for supply services – source, distribution and outlets. Because of the dual nature of air handling installations, involving extract and supply, the 'outlets' heading requires to be revised to 'outlets/inlets'. Examples of the respective components for air handling are:

Source	**Distribution**	**Outlets/Inlets**
Heating and cooling batteries	Ductlines	Grilles
Fans		Ventilators
Boiler plant		Terminals

Measurement Rules

Air handling installations are covered by SMM7 Appendix B/Section U Ventilation/Air conditioning systems, or Sections T40/41 Warm air heating for the classification of the installation and Rules Y30–59 for the detailed rules of measurement.

'**Source**' items are mainly enumerated under SMM7 Rules Y40–46 General air ductline equipment. These are defined in the Measurement Code (page 42) which lists such items as fans, filters, heating and cooling batteries, humidifiers and package air handling units. The actual classifications within Rules Y40–46 are listed in the Detailed Contents of SMM7 on page 7 and these references are used in the Commentary columns of the Worked Examples which follow.

'**Distribution**' items are invariably ductlines which are measured under SMM7 Rules Y30/31 Air ductlines and Air ductline ancillaries. Ducting is measured by length with fittings, special joints and the like enumerated as 'extra over the pipes in which they occur' as Y30.1 and Y30.2. Examples of duct fittings which are 'extra over' are given in the Measurement Code (page 42) and include such items as stop ends, bends, offsets, diminishing pieces and junctions. Ducting ancillaries, on the other hand, are enumerated *full value* as Y31.4, and these are also defined in the Measurement Code as such items as grilles, diffusers, dampers and shutters. Should these items occur at the end of a duct, then refer to the paragraph on 'Outlets/Inlets' below.

It is obviously necessary to differentiate between fittings and ancillaries in order to itemise them correctly and there is a requirement to deduct or omit any significant length of duct displaced by a full value ancillary.

It will be evident that the rules of measurement for ducts are very similar to those for pipelines. On a practical level of taking-off quantities, however, there is one significant difference which should be noted. When measuring pipelines it is usual to adopt a 'centre line of pipe' approach so that at junctions the branch length would be taken from the centre line of the main pipe. In the case of ductlines this approach would result in an unacceptable over-measurement because of the often quite large sectional size of ducts. It is therefore normal to measure branch ducts starting from the face of the main duct, as shown in the following diagram:

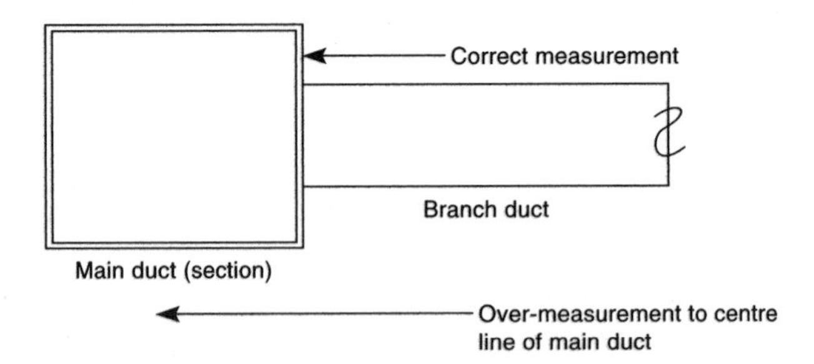

'**Outlets/Inlets**' comprise fairly simple components, owing to the nature of air handling installations such as grilles, terminals and the like. These are enumerated *full value* as ductline ancillaries under SMM7 Rules Y31.4 being defined in the Measurement Code (page 42).

As noted in the 'Distribution' paragraph above, some ductline ancillaries are within the ducting, forming part of the distribution system, whereas others are at the ends of the system and are outlets or inlets. All ductline ancillaries should be billed together under an appropriate heading ('Distribution' and 'Outlets/Inlets' being purely a convenient means used within this book to explain the general theory of building services measurement).

Special Locations of Work

When measuring ducting there is a requirement in SMM7 Rule Y30/31 M2 to keep work in plant rooms separate, which is in line with other sections of mechanical services. It should be noted however that, unlike pipelines, there are no other special locational requirements in SMM7 for ducting. The likely explanation is that ducting is usually in relatively awkward locations in any case, and the normal cost of execution will cover this feature.

Air handling plant is very likely to be housed within special plant rooms and should be identified separately under SMM7 Rule Y20/25 M2.

Taking Off Ductlines

Ductlines tend to be less numerous and complex than pipelines and consequently may often be successfully measured directly from the drawings, working logically along each line of ducting. Should the installation be more complex, then the surveyor can adopt the 'abstract' approach using the same technique as explained for pipelines.

WORKED EXAMPLE

The following **Worked Example 6/1** will demonstrate the application of the various Rules from SMM7 to typical air handling installations. Readers should refer to SMM7 and the Measurement Code while following this example, which will be cross-referenced where appropriate.

Worked example 6/1 is based on a simple air extract installation of part of a hotel, which covers all the measurement points without the complexities of a project that incorporates full air conditioning. It includes rigid and flexible ductlines with an extract fan and cowl.

This worked example is set out in draft bill format, which clarifies not only the SMM7 measurement requirements but also the various ancillary matters such as testing and sundries which may not appear clearly in a basic take-off.

Worked Example 6/1: Holiday Hotel – Ventilation

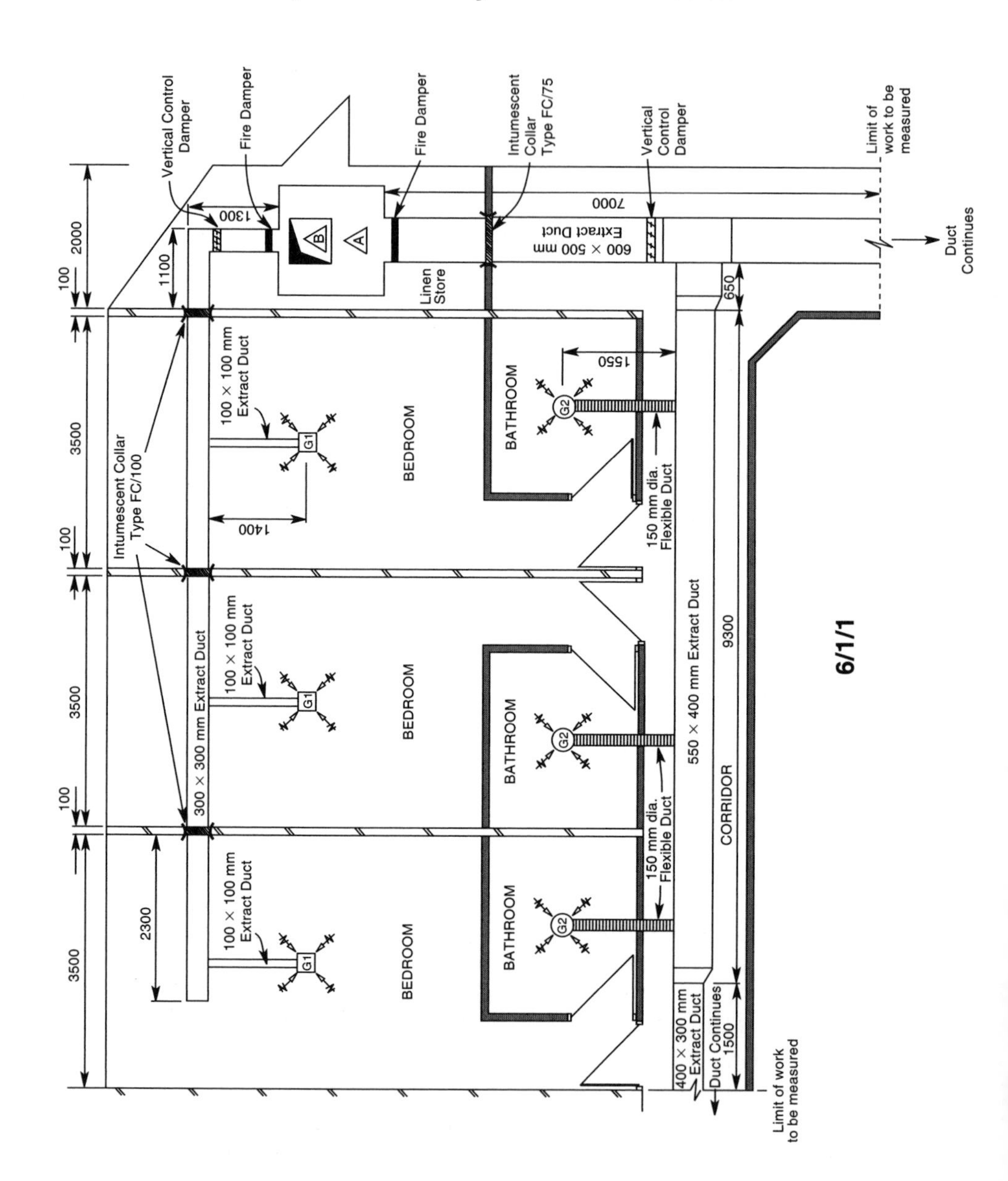

6/1/1

Worked Example 6/1: Holiday Hotel – Ventilation

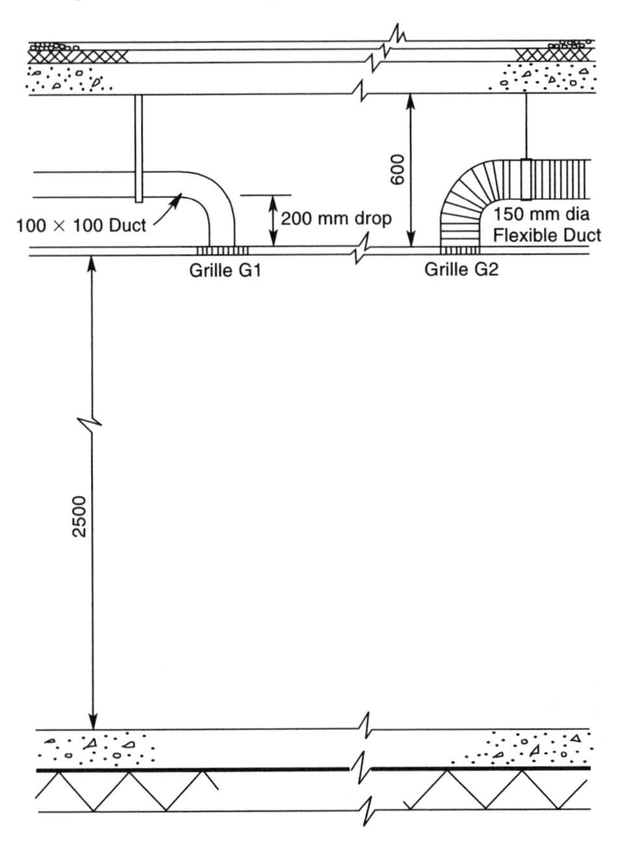

TYPICAL CROSS-SECTIONS

NOTES

1. Rectangular ductwork and fittings to be of hot dipped galvanised sheet to BS 2989 Grade Z2; flanged joints, fixed with prefabricated supports at 500 centres.
2. Flexible circular ductwork to be of tear resistant glass fabric with bonded galvanised spring steel helix; screw joints, fixed with proprietary supports at 400 centres.

KEY:

 'Compact' combination A.H.U. Type 3, with silencer, extract fan, anti-vibration mounting to concrete, extract fan duty 1.45 m^3/s at 185 Pa, suitable for 240 V/50 Hz electrical supply.

 'Roof Unit' low profile roof cowl Type RC/4.

 200 × 200 air diffusion grille; aluminium.

 150 diameter grille; PVC.

WORKED EXAMPLE 6/1			SHEET NR 1	
Commentary	Item Nr	Description	Unit	Qty
Project title		*HOLIDAY HOTEL*		
Sequential order of bills		*BILL NR 11*		
Common Arrangement of Work Section heading		*U – VENTILATION/AIR CONDITIONING SYSTEM*		
Heading from Appendix B.		***U10 General Extract***		
		Information provided		
This note covers the requirements of Rule P1 of sections Y30–59. Placing this first avoids repetition in each subsection of the bill.		The following represents the installation of an air extraction system within the bedroom and bathroom areas of the single storey hotel as Plans 6/1/1 & 2.		
Essential heading to indicate rules adopted for measurements following.		***Y30/31 Air Ductlines and Air Ductline Ancillaries***		
Y30.1.*.1		*GMS ducting, jointed with screwed flanges, supported with prefabricated MS brackets at 500 mm centres plugged and screwed to concrete soffit; all as specification clauses Y30/31/12*		
Y30.1.1.1.1 – Different sizes in sub-items. Rule Y30–M3		Ducting, straight, rectangular		
	1/1	100 × 100 mm	m	5
		Bedrooms 3/1.60		
Note: Linen store assumed *not* to be a 'plant room' as Y30/31–M2	1/2	300 × 300 mm	m	12
		Linen – bedrooms 10.70 Fan – corner (net) 1.00		
	1/3	400 × 300 mm	m	2
Example represents only sample part of whole hotel project.		Corridor – limit 1.50		
	1/4	550 × 400 mm	m	9
		Corridor 9.30		
	1/5	600 × 500 mm	m	8
		Fan – limit 7.00 Branch 0.65		
Y30.2.***		*Items extra over GMS ducting in which they occur*		
Y30.2.3.1	1/6	Fitting, stop end 300 × 300 duct	nr	1
		Bedroom 1		

WORKED EXAMPLE 6/1		SHEET NR 2		
Commentary	Item Nr	Description	Unit	Qty
Y30.2.*** (cont.)		*Items extra over GMS ducting in which they occur (cont.)*		
Y30.2.3.1	2/1	Fittings, radiused right-angle bend 100 × 100 duct	nr	3
		Bedrooms 3		
Not radiused and cheaper to make.	2/2	Fitting, right-angle 300 × 300 duct	nr	1
		Linen store 1		
	2/3	Fitting, diminishing piece 550 × 400 to 400 × 300	nr	1
		Corridor 1		
	2/4	Fitting, diminishing piece 600 × 500 to 550 × 400	nr	1
		Corridor 1		
	2/5	Fittings, right-angle junctions 300 × 300 to 100 × 100	nr	3
		Bedrooms 3		
	2/6	Fittings, right-angle junctions 550 × 400 to 150 diameter circular duct	nr	3
		Bathrooms 3		
	2/7	Fitting, right-angle junction 600 × 500 to 600 × 500	nr	1
		Corridor 1		
Y30.2.2.1.1		Special joints between duct and equipment (extract fan) with bolted flanges and neoprene gasket joints, reduced as required		
	2/8	300 × 300 duct	nr	1
		1		
	2/9	600 × 500 duct	nr	1
		1		
Y31.4.1.1		*Ancillaries to GMS ducting*		
Measurement Code page 42 for definition of ancillaries.	2/10	Air grilles 200 × 200 mm aluminium diffuser pattern to 100 × 100 duct	nr	3
		Bedrooms 3		

WORKED EXAMPLE 6/1			SHEET NR 3	
Commentary	Item Nr	Description	Unit	Qty
Y31.4.1.1 (cont.)		*Ancillaries to GMS ducting (cont.)*		
		Fire dampers as specification clause Y31/23		
	3/1	300 × 300 duct	nr	1
		Near fan 1		
	3/2	600 × 500 duct	nr	1
		Near fan 1		
		Vertical control dampers as specification clause Y31/24		
	3/3	300 × 300 duct	nr	1
		Linen store 1		
	3/4	600 × 500 duct	nr	1
		Corridor 1		
Y.31.7.1.1.2 Length refers to thickness of wall through which sleeve is to pass.	3/5	Intumescent collars type FC/ 100, length not exceeding 300 mm, as specification clause Y31/29; for 300 × 300 duct	nr	3
		Bedroom walls 3		
Y31.7.1.1.2	3/6	Intumescent collar type FC/75, length not exceeding 300 mm, as specification clause Y31/33; for 600 × 500 duct	nr	1
		Linen store wall 1		
Y30.1.*.1		*Glass fabric tear resistant circular ducting, with bonded reinforcing helix and screw joints, supported with proprietary brackets at 400 mm centres plugged and screwed to concrete soffit; all as specification clauses Y30/31/14*		
Y30.1.5.1.1	3/7	Ducting, flexible, circular; 150 mm diameter	m	5
		Bathrooms 3/1.70		
Y30.2.2.1 Rule D2	3/8	Extra over 150 diameter duct for special joints to circular spigot on GMS ducting with screwed clamp	nr	3
		Bathrooms 3		
Y31.4.1.1 Grilles require support as duct flexes.	3/9	Ancillaries to 150 diameter duct, air grilles 150 mm diameter PVC, including metal support collar screwed to hung ceiling frame	nr	3
		Bathrooms 3		

WORKED EXAMPLE 6/1			SHEET NR 4	
Commentary	Item Nr	Description	Unit	Qty
		Y40/46 General Air Ductline Equipment		
Y41.1.1.1.4 + 6	4/1	Extract fan A.H.U. type 3, 1.45 m^3/s with silencer and anti-vibration mounting bolted to concrete soffit, as specification clause Y40/2. (Electrical connections measured elsewhere.)	nr	1
		Linen store 1		
Y41.2.1.1	4/2	Roof cowl, low profile type RC/4 as specification clause Y40/10, set and bolted on concrete roof with spigot directly connected to above detailed fan	nr	1
		1		
		Y51–59 Testing and Sundries		
Y59.1.1	4/3	Marking positions of holes, mortices and chases in the structure for the whole of the foregoing General Extract system	item	
Y51.4.1	4/4	Testing and commissioning the whole of the foregoing General Extract system as per specification clause Y51/3	item	
P31 Rule M2		***P31 Holes, Chases, Supports for General Extract System***		
P31.20 Cutting or forming holes for services installations are enumerated, see Chapter 2 – 'Builder's Work'.	4/5	Cutting or forming holes for circular ducts girth not exceeding 1.00 m through concrete blockwork wall 100 mm thick and making good	nr	3
		Bathrooms 3		
	4/6	Cutting or forming holes for rectangular ducts incorporating intumescent collars total girth not exceeding 1.00 m through brick wall half brick thick and making good	nr	3
		Bedrooms 3		
	4/7	Cutting or forming holes for rectangular ducts incorporating intumescent collars total girth over 2.00 m but not exceeding 3.00 m concrete blockwork wall 100 mm thick and making good	nr	1
Hole in *in situ* concrete roof for fan would be measured as formwork within the concrete work section, see Chapter 2 – 'Builder's Work'.		Linen store 1		

7 Measurement of Electrical Services

Introduction

In Chapter 2 the general approach to the billing and measurement of building services was covered, indicating three basic components of supply services installations – source, distribution and outlets. In the specific case of Electrical Services the respective components are:

Source	**Distribution**	**Outlets**
Main switchgear and distribution boards	Cables and conduit Ducting or cable tray	Lighting points Power outlets or other outlets

Bills of quantities for smaller installations fully comply with the 'Source, Distribution and Outlets' concept but larger installations have further levels of distribution necessary to handle the larger electrical currents involved. However, the basic concept is still relevant to these further levels of distribution.

The electrical energy is generally provided by an electrical supply authority, which feeds power to the building's main switchgear via a metering system to measure the use of electricity for charging purposes. The incoming mains cabling, metering and connections to client's switchgear are usually provided by the electricity supply authority and would normally be included in a bill of quantities as a Provisional Sum (SMM7 rule A53 and general rule 10). Some specialist electrical installations include generating equipment to provide the basic electrical energy. This could apply in remote locations where no suitable electrical mains supply is available, or more likely the generating capacity would be provided as a standby source in the event of failure in the public electricity supply. Installations which are often provided with emergency generators occur in hospitals, where continuity of supply is essential to life support equipment for patients.

The voltage of the mains supply throughout the United Kingdom has been brought into line with that of continental Europe since January 1995. Smaller installations such as individual dwellings, small commercial buildings and small shops are usually supplied with single-phase AC 50 Hz at 230 volts nominal. Larger installations are usually supplied with three-phase AC 50 Hz at 400 volts nominal.

Measurement Rules

'**Source**' items are billed under the classification V20 LV Distribution (SMM7 Appendix B) and enumerated under Y71 LV Switchgear and distribution boards. Typical items in this category would be main switches, distribution boards and, in smaller installations, consumer units. These items are proprietary manufactured units and tend to be specified by reference to makers' catalogues. This is necessitated by the wide range of qualities and prices in which these units are available. Other features within the V20 classification of the bill would include associated items for Y60 Conduit and cable trunking, Y61 HV/LV Cables and wiring, Y62 Busbar trunking and Y80 Earthing and bonding components.

'**Distribution**' items are billed under various classifications in Section V of SMM7 Appendix B according to their purpose, such as V21 General lighting, V22 General LV Power and V40 Emergency lighting. It should be noted that classification V90 General lighting and power (small scale) allows combination billing for simple installations, but in practice this approach is not considered to benefit clarity and is therefore not included here. All 'Distribution' classifications by their nature include cabling, protection systems (conduits, cable trunking, cable tray) and earthing. These items may be measured in detail by the linear metre – see SMM7 Rules Y60.1, Y60.5 and Y61.1 as examples. In order to measure in this way (complete with methods of fixing and backgrounds stated) it is necessary to know in detail where cable runs are located. This information is rarely known for other than the larger cabling involved in sub-mains distribution, which in practice is often the only type of cabling that is measured in detail by length.

Cabling involved in final lighting and power circuits is not normally designed in detail, and the actual routes and locations of cable runs are usually left to the contractor who in turn will often leave this to site supervisory staff to decide. This situation makes detailed measurement of final circuits very difficult for the quantity surveyor, who would require a very intimate knowledge of the installation in order to make an educated estimate of the linear metres involved from layout drawings. This detailed requirement pertained in SMM5, but in practice many surveyors avoided the problem completely by putting the whole installation of electrical works into the bill as a Provisional Sum. SMM7, however, has recognised that the trade tends to price final circuits on a 'points' basis and has incorporated Rule Y61.19 (Cables and conduit in final circuits) which reflects this practical approach.

Measurement on a Points Basis
Enumeration on a points basis under Rule Y61.19 is restricted to final circuits of a domestic or similar simple installation from distribution boards (SMM7, Y61 Rules M6 & M7), otherwise detailed measurement is required. Further valuable guidance on Rule Y61.19 is given in the SMM7 Measurement Code, but the actual decision on whether to enumerate points or to measure in detail is wisely left to the individual surveyor dealing with a particular situation. In practice the largest cabling installed in final circuits of domestic premises is 10 mm^2 copper conductors for 7 to 9 kW instantaneous showers with a route length of more than 25 linear metres. Most commonly this distance will be less, and the voltage drop considerations will allow 6 mm^2 copper conductors to be used for the majority of shower circuits. Additionally, as domestic cooker installations will not normally exceed 6 mm^2 copper conductor size, it is reasonable to assume that this size of cable and rating of circuit are good practical maximum values for measurement on a points basis.

To conclude on the application of the very useful Rule Y61.19, in practical circumstances it may be summarised as follows:

Domestic final circuits:	All circuits normally enumerated on a points basis.
Non-domestic final circuits:	(i) Simple single-phase circuits enumerated on a points basis up to 6 mm^2 copper conductor size or equivalent. (ii) All other circuits measured in detail by metres with appropriate labour items.

'**Outlets**' items are billed under the same classification headings as their associated distribution final circuits; for example, the heading for V21 General lighting would include cabling and conduit along with the outlets. Outlets in the case of lighting would comprise items measured under Y73 Luminaires and lamps plus items under Y74 Accessories for electrical services. These items are all enumerated and include any associated conduit box stating background and fixings.

WORKED EXAMPLES

The following worked examples 7/1 and 7/2 will demonstrate the application of the various Rules from SMM7 to typical installations. Readers should refer to SMM7 and the Measurement Code while following these examples, which will be cross-referenced as appropriate.

These worked examples are set out in draft bill format, which clarifies not only the SMM7 measurement requirements but also the various ancillary matters such as testing and sundries which may not appear clearly in a basic take-off.

Worked Example 7/1

This example is based on a single shop building representing a very straightforward installation, but which nevertheless covers all the essentials for the preparation of a bill of quantities and measurement of electrical work.

Worked Example 7/2

This example is more advanced and includes the additional work of an underground main cable to serve a new detached building on a client's property. It also contains a final circuit which has been measured in detail as opposed to enumeration by points.

Worked Example 7/1: Shop Unit – Electrical Installation

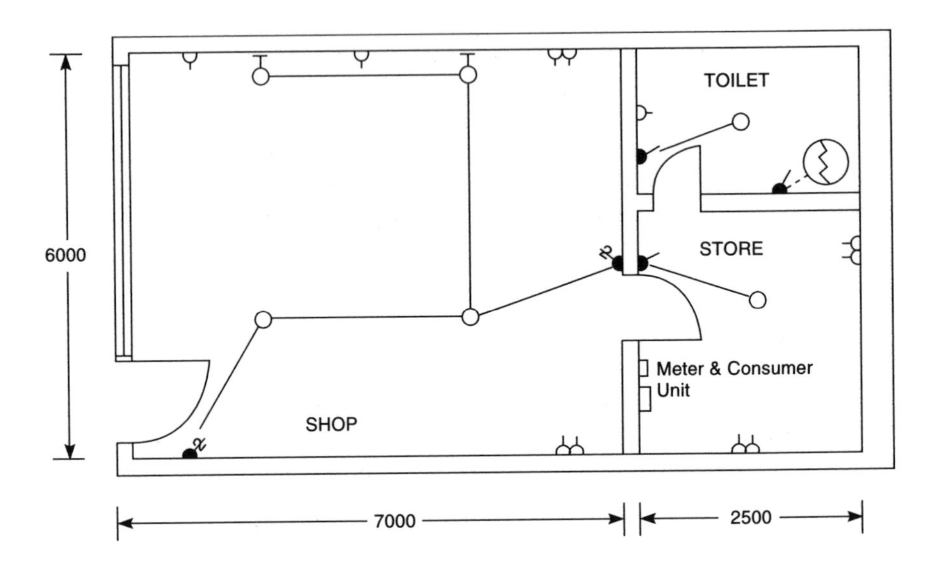

Construction

Walls	Brick and concrete block throughout.
Floors	Concrete with cement and sand screed.
Roof	Timber joisted flat construction.
Height	Finished floor to ceiling height – 2700 mm.

Specification

The following represents a brief specification to indicate the general requirements for the installation but in order to price work in the commercial world a fully detailed specification would be prepared and issued with the Bill. Such a detailed document is beyond the scope of this book but references to a hypothetical full specification are made in some of the items in the Worked Example where this would happen in practice.

Voltage	230 volt 50 Hz AC mains provided by public electricity authority.
Enclosure	Heavy gauge mild steel screwed conduit which acts as earth continuity.
Cable	PVC insulated single core cables drawn into conduit and colour coded as appropriate.
Lighting	To be wired in two 5 amp circuits in 1 mm^2 cables.
Power	To be wired in two 30 amp ring circuits in 2.5 mm^2 cables.
Water Heater	To be wired in one 15 amp radial circuit in 2.5 mm^2 cables.
Consumer Unit	Surface type metal clad with 60 amp main switch and 6 ways (5 per distribution sheet and 1 spare way).
Main Cable	To be wired in 10 mm^2 cables protected by short length of trunking between meter and consumer unit.
Accessories	Switches, ceiling roses and power outlets to be white plastic pattern.
Regulations	The installation will comply with the latest IEE Regulations for Electrical Installations: Regulations for the Electrical Equipment of Buidings.

WORKED EXAMPLE 7/1 – SHOP UNIT

Electrical Installation – Distribution Sheet

Location	Lighting							Power			Water Heater		Remarks
	Circuit Nr	Points		Switches		Lamps		Circuit Nr	13 Amp Single SSO	13 Amp Double SSO	Circuit Nr	15 Amp Spur	
		Type	Nr	One way	Two way	Nr	Watts						
Shop	1	Ceiling*	2		2	2	60*	3	2	2			* Circular Fluorescent
		Wall	2			2	150						
Toilet	2	Ceiling	1	1		1	100	4	1		5	1	
Store	2	Ceiling	1	1		1	100	4		2			
Totals			6	2	2	6			3	4		1	

WORKED EXAMPLE 7/1		SHEET NR 1		
Commentary	**Item Nr**	**Description**	**Unit**	**Qty**
		SHOP UNIT 7/1		
Sequential order of bills.		*BILL NR 11*		
Common Arrangement of Work Section heading from SMM7 Appendix B.		*V – ELECTRICAL SUPPLY/ POWER LIGHTING SYSTEMS*		
		Information provided		
This note covers the requirements of Rule P1 of sections Y60–92 and Rule P2 of Y61. Placing this first removes the need for repetition within various subsections of the bill.		The following represents the electrical installation of 1 shop unit as detailed on Plan 7/1/1 comprising lighting, power, water heating and main switchgear. A distribution sheet is given with Plan 7/1/1.		
Heading from Appendix B.		***V20 LV Distribution***		
Essential heading to indicate the rules adopted for the measurements following.		***Y60 Cable Trunking***		
Y60.5.1.1 Trunking not always installed – see commentary to item nr 1/4. Y60/M4, M5, C5	1/1	Cable trunking (straight) 40 × 40 section having lid with screw fixings all in enamelled MS; plugged and screwed to masonry 1.00	m	1
Y60.6.1.1 Y60/C6	1/2	Extra over 40 × 40 trunking for ends 2	nr	2
Y60.7.3.1 Y60/C6	1/3	Connection of cable trunking to control gear with 30 × 30 flanges and forming holes consumer unit 1 meter 1	nr	2
Y61 heading HV does not apply therefore omitted.		***Y61 LV Cables and Wiring***		
Y61.1.1.1 Y61/M2, C3 If trunking omitted then *insulated and protected cables* required for mains 'tails'.	1/4	10 mm^2 single core PVC insulated colour coded cable laid into trunking line & neutral 2/1.00 M3 allowances 2/0.60 (cons. unit) 2/0.60 (meter tails)	m	4

WORKED EXAMPLE 7/1		SHEET NR 2		
Commentary	Item Nr	Description	Unit	Qty
IEE requirement		***Y80 Earthing***		
Y61/1.1.1	2/1	10 mm^2 single core PVC insulated colour coded cable laid into trunking	m	2
		1.00 M3 allowance 0.60		
Y61.1.1.3	2/2	10 mm^2 cable as last but fixed to masonry surfaces with screwed and plugged cable clips	m	1
Open earth cable clamp on surface of wire armour of mains cable all provided by the electricity supply authority. Connecting tail of cable to clamp deemed to be included Rule Y61 C3(c).		0.60 M3 allowance nil (does not enter)		
		Y71 LV Switchgear and Distribution Boards		
Y71.1.1.1 Y71.1.1.3	2/3	Consumer Unit as specification clause Y71/8 surface type metal clad with 60 amp main switch, (6) HRC fuseways comprising: 2 × 5 amp; 2 × 30 amp; 1 × 15 amp; and 1 × 30 amp spare way; unit plugged and screwed to masonry	nr	1
		1		
Y61.19 Measurement Code page 43 Rule 19. Y61 Rule P2(a) & (b) requirements are on the Drawings.		***V21 General Lighting*** ***Y61 Conduit and Cables in Final Circuits*** *Final circuits in heavy gauge MS conduit with single core 1 mm^2 PVC insulated and colour coded cables (230 volt, 5 amp lighting circuits) generally concealed to backgrounds comprising timber roof and plastered masonry walls*		
Y61.19.2.3	2/4	Lighting outlets 6	nr	6
Y61.19.2.4	2/5	One way switches 2	nr	2
Y61.19.2.5	2/6	Two way switches 2	nr	2

WORKED EXAMPLE 7/1			SHEET NR 3	
Commentary	Item Nr	Description	Unit	Qty
		Y73 Luminaires and Lamps		
Y73.2.2.1 BC is abbreviation for bayonet connection.	3/1	Plain pendants consisting of 0.75 mm^2 three core PVC insulated and protected flexible drop not exceeding 1.00 m, white plastic rose and brass BC lamp holder with shade ring, include round CI conduit box screwed to timber	nr	2
		store/toilet 2		
Y73.2.1.1	3/2	Circular fluorescent ceiling mounting fittings with pearl globe as specification clause V21/11 complete with 60 watt tube, include for round CI conduit box screwed to timber	nr	2
		shop 2		
Y73.2.1.1 BC is abbreviation for bayonet connection.	3/3	Wall bracket fittings with pearl shades and BC lamp holders as specification clause V21/12, include for round CI conduit box screwed to masonry	nr	2
		shop 2		
Y73.3.1 BC is abbreviation for bayonet connection.		BC coiled filament pearl lamps		
	3/4	100 watt	nr	2
		store/toilet 2		
	3/5	150 watt	nr	2
		shop 2		
		Y74 Accessories		
Y74.5.1.1		5 amp single pole silent action white plastic plate switches, include steel conduit box plugged and screwed to masonry		
	3/6	One way	nr	2
		store/toilet 2		
	3/7	Two way	nr	2
		shop 2		

WORKED EXAMPLE 7/1		SHEET NR 4		
Commentary	**Item Nr**	**Description**	**Unit**	**Qty**
		V22 General LV Power		
		Y61 Cables and Conduit in Final Circuits		
Y61.19.2 Measurement Code page 43 Rule 19.		*Final circuits in heavy gauge MS conduit with single core 2.5 mm² PVC insulated and colour coded cables, generally concealed in backgrounds comprising plastered masonry walls and screeded concrete floors*		
Y61.19 Rule S8 SSO is abbreviation for switched socket outlet.	4/1	Switch sockets 230 volt 30 amp ring circuits	nr	7
		single SSO 3 double SSO 4		
Different type of circuit requiring separate item refer to Measurement Code P43 Rule 19.	4/2	Immersion heater 230 volt 15 amp single outlet radial circuit	nr	1
		toilet 1		
Y61.2.1.1	4/3	Flexible cable connection between control switch and connection box on 3000 watt unit water heater, comprising three core 50/0.25 mm² butyl rubber insulated and sheathed flexible, length not exceeding 1.00 m	nr	1
		Y74 Accessories		
Y74.5.1.1	4/4	13 amp switched single socket outlets, white plastic plate pattern, include appropriate steel conduit box plugged and screwed to masonry having 1 mm² copper earth connector with PVC colour coded sleeve between box and outlet	nr	3
		from dist sheet 3		
Y74.5.1.1	4/5	13 amp switched double socket outlets otherwise as item 4/4 above	nr	4
		from dist sheet 4		
Y74.5.1.1	4/6	Water heater control switch double pole 15 amp rating, white plastic plate pattern embossed 'WATER HEATER' with warning neon and outlet for flexible cord, include conduit box with earth connector as described in item 4/4 above	nr	1
		from dist sheet 1		

WORKED EXAMPLE 7/1			SHEET NR 5	
Commentary	Item Nr	Description	Unit	Qty
This section comprises three SMM7 headings which are best combined.		***Y81, 82, 89 Testing, Identification and Sundries*** N.B. Items in this section of the Bill refer to the whole of the foregoing installation.		
Y81.1.1 refers to bonding of non-electrical installation metal-work which requires to be tested and bonded to earth. SMM7 requires this work to be by a Provisional Sum but in this very simple project it is not unreasonable to expect the contractor to price as a lump sum.	5/1	Additional bonding to extraneous metal in accordance with the IEE Regulations to be priced on a lump sum basis to cover the bonding required for the plumbing and heating installations and the roller shutter to the shop front	Item	
Y89.2 This item should always be included as it affects every installation.	5/2	Marking position of holes, mortices and chases in the structure	Item	
Y82.4 comprises identity plates, discs, etc. which would be unlikely in small jobs. However a possible example is given.	5/3	Identification diagram of circuit system drawn in ink on paper on hardboard about 300 × 200 mm coated in clear paper varnish plugged and screwed to masonry	nr	1
Y81.5.1 This item should always be included as every installation would normally be tested.	5/4	Testing and commissioning the whole installation in accordance with the IEE Regulations as specification clause Y82/14	Item	
This work is often referred to as 'Builder's work in connection with electrical installation' and may be billed elsewhere in the Bill at the surveyor's discretion. However it must be included somewhere in the Bill as it covers all preparatory work and making good to the structure and finishes. P31.19 measures this on a points basis with associated switches deemed included (Rule P31 C3). The quantities are obtained from the distribution sheet.		***P31 Holes and Chases for Electrical Installation*** N.B. Items in this section of the Bill refer to the whole of the foregoing installation. *Cutting or forming holes, mortices, sinkings and chases for electrical installation comprising concealed MS conduits and make good*		
P31.19.1.1	5/5	Luminaire points	nr	6
P31.19.1.2	5/6	Socket outlet points	nr	7
P31.19.1.3 (Water heater)	5/7	Fitting outlet point	nr	1
P31.19.1.4 (Consumer unit)	5/8	Control gear point	nr	1

Worked Example 7/2: Gate House – Electrical Installation

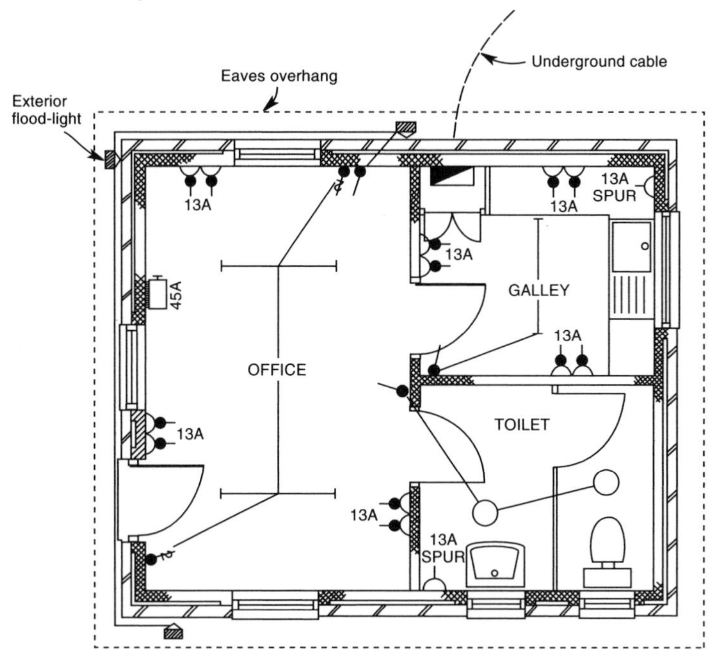

PROPOSED GATE HOUSE – ELECTRICAL LAYOUT

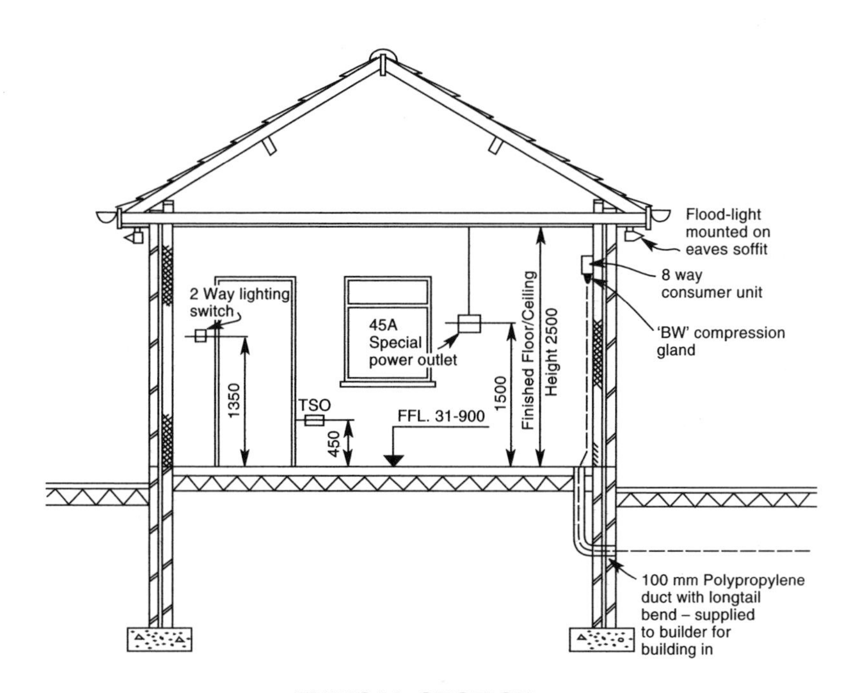

TYPICAL SECTION

7/2/1

Worked Example 7/2: Gate House – Electrical Installation

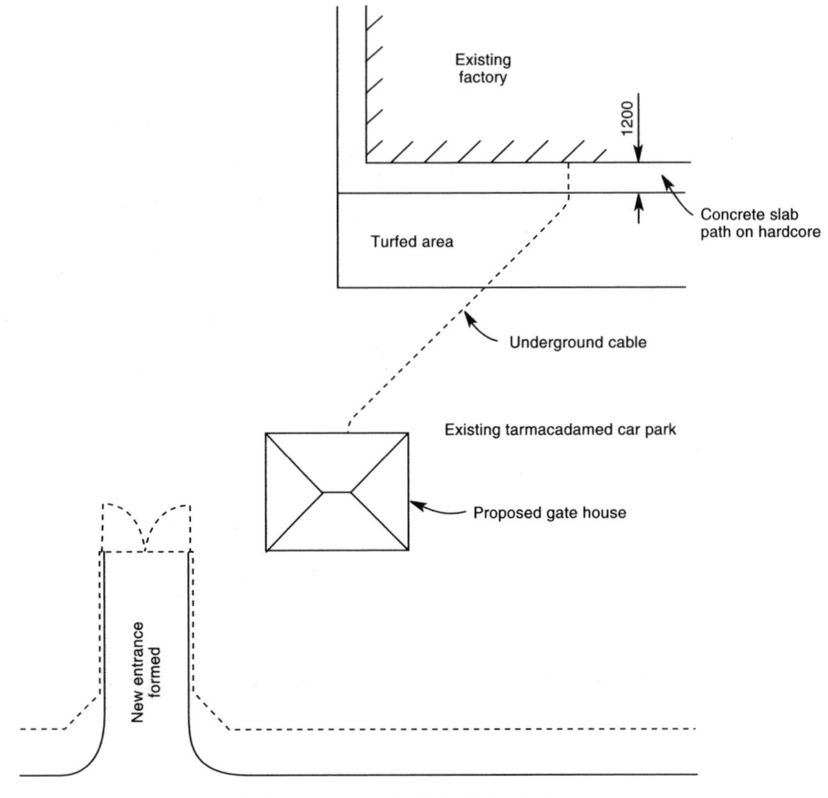

GENERAL LAYOUT PLAN

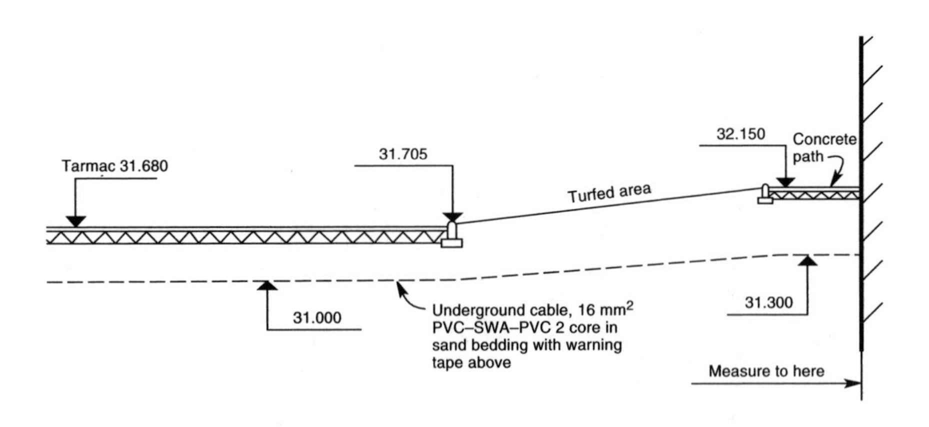

CROSS-SECTION OF U/G CABLE

7/2/2

Worked Example 7/2: Gate House – Electrical Installation

DISTRIBUTION SHEET										
LOCATION	LIGHTING					POWER				REMARKS
	Circuit No.	Points	Switches		Lamps	Circuit No.	13A TSO	13A Spur	Special Power	
			1 Way	2 Way						
Office	1	2(F)		2	60 W(F)	4 6	3		 1	(F) 1200 mm Fluorescent 45 Amp Metalclad switch
Toilet	2	2(S)	1		75 W	5		1		(S) Soffit mounted fitting
Galley	2	1(F)	1		60 W(F)	5	3	1		(F) 1200 mm Fluorescent
External	3	3(X)	1		150 W(X)					(X) 150 Watt Sodium combination flood-lights
Totals	–	8	3	2	–	–	6	2	1	

SPECIFICATION NOTES

Generally – All work to comply with IEE regulations.

Consumer unit – Metal clad, surface pattern, 8 HRC fuseways, 6 circuits as per distribution sheet and 2 spare ways × 30 amp.

Mains cable – 2 core × 16 mm² PVC–SWA–PVC cable laid in trench and brought into gate house as shown on drawing, terminating at consumer unit with BW compression gland.

Lighting circuits – Generally in PVC insulated and protected cables, concealed system with lightweight conduit drops to switches in plastered blockwork walls. Cables – twin and triple 1.5 mm² conductors with 1 mm² ECC.

Power circuits – Generally in heavy gauge mild steel conduit concealed in screeded floors and plastered blockwork walls. Cables – 2.5 mm² PVC insulated colour coded. Special power wired in 10 mm² PVC insulated cables in 20 mm conduit.

Power outlets – Twin socket outlets to be switched white plastic plate pattern. 13 amp switched single spur outlets to have cable outlet and fuse carrier. No plugs to be provided.

Special power outlet – 45 amp switched appliance outlet, surface metal clad pattern with screwed outlet for flexible conduit connection by others.

Fluorescent lights – 1200 mm × 60 watt, white enamelled finish complete with tubes, Model No. 1200/2.

Soffit mounted light – Circular pattern with pearl plastic bowl cover, ceramic BC lamp holder Model No. 800/5.

Sodium flood-lights – 150 watt combination flood-lights complete with high pressure sodium lamps, soffit mounting Model No. 2000/1.

WORKED EXAMPLE 7/2			SHEET NR 1	
Commentary	Item Nr	Description	Unit	Qty
		GATE HOUSE 7/2		
Sequential order of bills.		*BILL NR 10*		
Common Arrangement of Work Section heading from SMM7 Appendix B.		*V – ELECTRICAL SUPPLY/ POWER LIGHTING SYSTEMS*		
		Information provided		
This note covers the requirements of Rule P1 of sections Y60–92 and Rule P2 of Y61. Placing this first removes the need for repetition within various subsections of the bill.		The following represents the electrical installation of 1 gate house as detailed on Plans 7/2/1 & 2 comprising lighting, power, underground main supply and main switchgear. A distribution sheet is shown on Plan 7/2/3.		
Heading from Appendix B.		***V20 LV Distribution***		
Essential heading to indicate the rules adopted for the measurements following.		***Y60 Conduit and Cable Trunking***		
Y60.1.1.1	1/1	Conduit (straight) 100 mm polypropylene supplied to builder for building in	m	1
		1.00		
Y60.1.2.1 adapted – enumeration more appropriate.	1/2	Conduit (curved) 100 mm polypropylene proprietary long tail bend supplied to builder for building in	nr	1
		1		
Y61 heading HV does not apply therefore omitted		***Y61 LV Cables and Wiring***		
		<u>*Cable 16 mm² two core PVC/SWA/PVC*</u>		
Y61.1.1.1 Y61/M2, C3	1/3	Drawn into conduit or ducts	m	2
		straight 1.00 bend 0.50		
Y61.1.1.3 steel wire armoured (SWA) cable on surface	1/4	Fixed to surface of masonry with plugged and screwed cleats at 500 mm centres	m	3
		gate house wall 2.00 M3 allowance 0.60 (cons.unit)		
Y61.1.1.5	1/5	Laid in trenches	m	15
		(scaled) 15.00		

WORKED EXAMPLE 7/2		SHEET NR 2		
Commentary	Item Nr	Description	Unit	Qty
Y61.3.1	2/1	Cable termination gland to metal clad consumer unit, BW compression pattern 1	nr	1
		Y80 Earthing		
Y80.14 adapted to suit cable earth instead of tape. Y80 Rules M5 & C5.	2/2	Earth connection between cable and consumer unit earth block with clamp on SWA and short length of 10 mm^2 single core PVC insulated colour coded cable 1	nr	1
		Y71 LV Switchgear and Distribution Boards		
Y71.1.1.1	2/3	Consumer unit as specification clause Y71/9 surface type metal clad with 60 amp main switch, 8 × HRC fuseways comprising: 3 × 5 amp; 2 × 30 amp; 1 × 45 amp; and 2 × 30 amp spare ways; unit plugged and screwed to masonry 1	nr	1
		V21 General Lighting		
		Y61 Conduit and Cables in Final Circuits		
Y61.19 Measurement Code page 43 Rule 19. Y61 Rule P2(a) & (b) requirements are on the Drawings.		*Final circuits in PVC insulated and protected cables – twin and triple 1.5 mm^2 as appropriate (230 volt, 5 amp lighting circuits) generally concealed in backgrounds comprising timber roof and plastered masonry walls – all drops to switches to be protected in light gauge steel conduit*		
Y61.19.1.3	2/4	Lighting outlets 5	nr	5
External work not required to be stated by SMM7 but there is a definite weather risk – hence item.	2/5	Lighting outlets at external eaves soffits 3	nr	3
Y61.19.1.4	2/6	One-way switches 3	nr	3
Y61.19.1.5	2/7	Two-way switches 2	nr	2

WORKED EXAMPLE 7/2			SHEET NR 3	
Commentary	Item Nr	Description	Unit	Qty
		Y73 Luminaires and Lamps		
Y73.2.1.1	3/1	Soffit mounted 1200 mm × 60 watt fluorescent fittings complete with tube, Model No. 1200/2, screwed to timber	nr	3
		3		
Y73.2.1.1 BC is abbreviation for bayonet connection.	3/2	Soffit mounted circular fittings with pearl plastic bowl cover, ceramic BC lamp holder, Model No. 800/5, screwed to timber	nr	2
		2		
Y73.2.1.1 'externally' included as workmanship must maintain the weather resistance of the unit.	3/3	Soffit mounted 150 watt combination flood-light complete with high pressure sodium lamps, Model No. 2000/1, screwed externally to timber	nr	3
		3		
Y73.3.1	3/4	BC coiled coil pearl lamps, 75 watt	nr	2
		toilet 2		
		Y74 Accessories		
Y74.5.1.1		5 amp single pole rocker action white plastic plate switches, include steel conduit box plugged and screwed to masonry		
	3/5	One way	nr	3
		3		
	3/6	Two way	nr	2
		2		

WORKED EXAMPLE 7/2		SHEET NR 4		
Commentary	Item Nr	Description	Unit	Qty
		V22 General LV Power		
		Y61 Cables and Conduit in Final Circuits		
Y61.19.2 Measurement Code page 43 Rule 19.		*Final circuits in heavy gauge MS conduit with single core 2.5 mm² PVC insulated and colour coded cables, generally concealed in backgrounds comprising plastered masonry walls and screeded concrete floors*		
Y61.19 Rule S8.	4/1	Sockets, switch sockets 230 volt 30 amp ring circuits	nr	8
SSO is abbreviation for switched socket outlet.		SSO 6 13 amp spurs 2		
		Y74 Accessories		
Y74.5.1.1	4/2	13 amp switched double socket outlets, white plastic plate pattern, including appropriate steel conduit box plugged and screwed to masonry having 1 mm² copper earth connector with PVC colour coded sleeve between box and outlet	nr	6
		from dist. sheet 6		
Y74.5.1.1	4/3	13 amp switched single spur outlets with outlet for flexible cord and fuse carrier, include conduit box with earth connector as last item	nr	2
		from dist. sheet 2		
Special heading to clearly signify differences with previous section – this work is cabled in 10 mm² conductors and is therefore measured in detail.		***V22 Special LV Power*** ***Y60 Conduit*** *Heavy gauge steel conduit with screwed fittings*		
Y60.1.1.1	4/4	Straight 20 mm diameter fixed to timber with screwed saddles	m	5
		in roof 2.00 3.20		
Y60.1.1.1	4/5	Straight 20 mm diameter fixed to masonry surfaces with plugged and screwed saddles	m	1
		in unplastered cupboard 0.60		

WORKED EXAMPLE 7/2		SHEET NR 5		
Commentary	Item Nr	Description	Unit	Qty
		Heavy gauge conduit (continued):		
Y60.1.1.1	5/1	Straight 20 mm diameter fixed in chases in masonry with driven crampets	m	1
		drop to outlet 1.10		
Y60.4.1.1 Along with Y60.3 these are special connections not to be confused with the deemed included conduit joints mentioned in Rule C3.	5/2	Connection of 20 mm diameter conduit to consumer unit include forming hole and providing lock nuts and star washer	nr	1
		1		
		Y61 LV Cables and Wiring		
		PVC Insulated and colour coded cables		
Y61.1.1.1 Y61 Rule M2. All quantities × 2 for line and neutral cables	5/3	10 mm^2 single core drawn into conduit	m	16
		length of item 4/4 × 2 = 2/5.20 length of item 4/5 × 2 = 2/0.60 length of item 5/1 × 2 = 2/1.10 allowances Y61 Rule M3 M3(a) switch 2/0.30 M3(b) consumer unit 2/0.60		
		Y74 Accessories		
Y74.5.1.1	5/4	45 amp switched fixed appliance outlet surface metal clad pattern with screwed outlet for flexible conduit (by others) plugged and screwed to masonry	nr	1
		office 1		
Refer to the commentary notes from Worked Example 7/1 – Y81, 82, 89 for explanation of this section.		***Y81, 82, 89 Testing, Identification and Sundries***		
		N.B. Items in this section of the Bill refer to the whole of the foregoing installation.		
Y81.1.1	5/5	Additional bonding to extraneous metal in accordance with the IEE Regulations to be priced on a lump sum basis to cover the bonding required for the plumbing and heating installations	Item	
Y89.2	5/6	Marking position of holes, mortices and chases in the structure	Item	
Y81.5.1	5/7	Testing and commissioning the whole installation in accordance with the IEE Regulations as specification clause Y81/2	Item	
Assumed that no identification or drawings required for this small installation.				

WORKED EXAMPLE 7/2		SHEET NR 6		
Commentary	Item Nr	Description	Unit	Qty
P30 Rule M2		***P30/31 Trenches, Holes, Chases for Electrical Installation***		
P30 Information Provided not covered at start of this Bill – therefore required here. As full details of ground water levels, trial pits etc. will be elsewhere it is enough to cross-reference in this case.		***Information Provided:*** for information regarding the nature of excavation work refer to the Bill Nr 2 – Groundwork.		
P30.1.1.2 (Note – for explanation of depths and runs, see Additional Commentary at end of worked example.) P30 Rules M3, M4, D2, C1	6/1	Excavating trenches to receive cables not exceeding 200 mm nominal size include bedding in sand at least 100 mm thick all round cable: average depth not exceeding 750 mm	m	15
Cable laid with two bends but this does not require 'curved' trench to be measured – large underground cables however involve laying to carefully controlled radii and these would need curved trench items.		whole route 15.00 (scaled)		
P30.2		*Items extra over excavating trenches, irrespective of depth*		
P30.2.2.2 Width determined by Rule Y30 M6 minimum 500 mm in this case.	6/2	For breaking out existing hard paving comprising concrete slabs on hardcore and neatly reinstate to match existing to specification clause Q25/16	m^2	1
		1.20 × 0.50		
P30.2.2.5	6/3	For breaking out existing hard paving comprising tarmacam on bottoming and neatly reinstate to match existing to specification clause Q22/18	m^2	4
		8.50 × 0.50		
P30.2.3.1	6/4	For lifting turf, preserving and neatly reinstate to match existing to specification clause Q30/19	m^2	3
		5.30 × 0.50		
P30.3.1 Always given as risk item – P30 Rule D4	6/5	Disposal, surface water	Item	
P30.13.1.1	6/6	Identification tape PVC strip incorporated into trench backfilling 300 mm above cable	m	15
		Quantity as item 6/1 = 15.00		

WORKED EXAMPLE 7/2		SHEET NR 7		
Commentary	Item Nr	Description	Unit	Qty
P31.19 covers all 'Builder's Work' required for electrical installations in new work situations but does not include work within existing buildings. This note is intended to make clear the extent of this Bill.		N.B. The following items refer to the whole of the foregoing installation for the Gate House. Work within the existing factory building does not form part of this Bill.		
P31.19.1 As there are two types of installation (both conduit and protected cable) in this project it is necessary to measure both types.		*Cutting and forming holes, mortices, sinkings and chases for electrical installation comprising concealed MS conduits and make good*		
P31.19.1.2	7/1	Socket outlet points 13 amp outlets 8	nr	8
P31.19.1.3	7/2	Fitting outlet point special power 1	nr	1
P31.19.1.4	7/3	Control gear point consumer unit 1	nr	1
P31.19.2 Lighting in concealed protected cable but with additional description to cover conduit drops to switch points which are deemed to be included – Rule P31/C3.		*Cutting and forming holes, mortices, sinkings and chases for electrical installation comprising concealed PVC protected cables with MS conduit drops to switches and make good*		
P31.19.2.1	7/4	Luminaire points 8	nr	8

Additional Commentary

Rule P30.1.** – Excavating trenches requires each 'run' of trench to be measured and categorised by average depth in not exceeding 250 mm stages. 'Run' is not defined in Section P but is defined for Drainage in R12/13 Rule D1 and it is reasonable to assume that the same considerations apply here. Therefore all of the trench from the existing factory wall to the gate house will constitute one 'run'. Having established the run it is now necessary to calculate the average depth of the trench and the more accurate average for this purpose is the weighted average. Commencing at the existing factory wall there is 1.2 m of trench under concrete paving of depth 850 mm, 5.3 m of trench under turf of average depth 778 mm and 8.5 m of trench under tarmac of depth 680 mm. The average weighted by the respective lengths works out at 728 mm which gives the category for the item of 'average depth not exceeding 750 mm'.

8 Measurement of Other Services

Introduction

The measurement of the commonly occurring types of building services is covered in the other chapters of this book, but there are several examples of building services which by their rarity or nature are seldom the subject of detailed measurement and billing. This chapter will consider the rules of measurement affecting those rare examples and will suggest suitable strategies for tendering for those examples where traditional measurement may not be appropriate.

Less Common Elements of Building Services

Steam Heating

The advantages of the added heat-carrying capacity of steam over water at normal pressure was recognised during the early developments of central heating systems, particularly in the case of larger buildings. However, there are serious disadvantages to the adoption of steam heating, namely higher installation and maintenance costs and possible operational noise through the system. More recent developments involving hot water heating at various pressures exceeding that of atmosphere have made steam heating virtually obsolete. However, it may still occur in projects where steam generation is required for other functions and also within alterations and extensions to existing steam heating installations.

The measurement of steam heating installations is similar to that of hot water heating with the added features of condensate pipelines and steam traps. Condensate pipelines are measured in the usual way (Rules Y10/11) but are kept separate from the measurement of steam pipes. Steam traps are measured as pipeline ancillaries Rule Y11.8.

Coal Fired Boilers

The relative popularity of the various fuels used in heating has varied over the years, mainly because of economic factors which in turn are affected by political and environmental considerations. Coal was the original fuel for the early central heating plants of the Victorian era, often hand fired by shifts of stokers. Modern coal fired installations use graded or pulverised fuel fed by automatic stokers or compressed air and can be highly automated.

Measurement of such plant from detailed drawings would be generally enumerated under Rules Y20–25 General Pipeline Equipment. Coal bunkers and waste ash bins are often constructed in concrete and would likely be taken within the Builder's Work section of the bill.

Oil Fired Boilers

Oil fired boilers were very popular in the 1960s because of the very competitive cost of fuel oil before the advent of natural gas. Oil is still used principally where mains gas is not available. Such installations require an oil storage tank installed over a catch pit to contain accidental leakage, and an oil fuel pipeline to the boiler provided with an automatic fuel cut-off valve which activates in the event of boiler malfunction. These pipelines and valves would be measured in the usual way under Rules Y10/11 Pipelines and Pipeline Ancillaries. Construction of tank houses and catch pits would likely be of brick and/or concrete and be measured within the Builder's Work section of the bill.

Electrical Power Generation

Power generation is provided in building projects for two reasons: location remote from mains electricity supply, or emergency plant to cover mains failure. The former is fairly rare in the UK but the latter is becoming more common with the continuity of power requirements of computer centres and life support apparatus in hospitals as examples.

SMM7 includes electricity generation plant in category V10 of Appendix B but does not specifically mention such work in the measurement rules Y70/71/72/92 covering switchgear, distribution boards, contactors, starters and motor drives. However, the same approach to the measurement of generation plant would be appropriate – enumeration with relevant details.

Communications/Security/Control Systems

The above are classified under Section W of Appendix B which includes such varied systems as W10 Telecommunications, W30 Data Transmission and W41 Security Detection and Alarms. As with other types of building services there is no reason why these installations should not be billed, provided that detailed drawings and specifications are available. Cables, accessories and equipment are measured in the same way as ordinary electrical installations – SMM7 Y60–92. In security systems the circuits are frequently wired on a radial basis and this should be stated in the item if the option to measure under Y61.19 is adopted – see also page 43 of the Measurement Code.

Contractor Designed Services

Sometimes the design of an installation is not fully detailed before tendering, with significant elements expected to be designed and provided by the contractor (usually based on a performance specification). It is quite difficult to incorporate this approach within a traditional bill of quantities under normal standard forms of contract, although the Measurement Code on page 13 suggests that General Rule 11 could be used to incorporate contractor designed work as measured work. The main difficulty is not in the measurement, but rather in attempting to provide the client with binding contractual performance guarantees for the design element.

Because of the foregoing reasons, contractor designed work is often made the subject of a Provisional Sum. This could be achieved either under SMM7 Rule A51 – if the nominated subcontractor provision is desired or under A54 – if a more general provisional work approach is considered more appropriate. General Rule 10 explains 'defined' and 'undefined' work covered by Provisional Sums, but in most cases contractor designed services should be capable of being defined.

Rule A51.1.1.1 mentions giving a prime cost or pc sum, but also requires the description stated in accordance with General Rule 10.3 which is all about Provisional Sums. To avoid possible confusion it is advisable to use the terminology given in the General Rules, which is *Provisional Sums.*

Transport Systems

Transport systems are included in SMM7 Section X covering such installations as lifts, escalators, moving pavements and pneumatic document handling. Unusually, the list of the various types of installation is contained in Measurement Rule M2 of Section X rather than in an appendix such as Appendix A or B.

Although the types of system within Section X vary considerably, there are certain features which frequently apply:

(1) an element of contractor design;
(2) patent design features unique to a particular manufacturer;
(3) a requirement for future maintenance of the installation, which in some cases is a legal safety requirement.

Any or all of these features make the traditional bill of quantities approach to procurement unlikely and the adoption of a Provisional Sum as discussed in the previous section more likely. The legal implications associated with the contractual enforcement of performance related design elements and of long-term performance related maintenance are relatively difficult to incorporate in traditional measured tendering documentation.

Additionally, the detailed design requirements imposed on the building structure by an individual manufacturer's product often demand that an early choice of supplier is made to enable drawings to be submitted in time for regulatory approval. This would be particularly true for installations such as lifts and escalators, which are often within the design zone for the emergency staircases and other means of escape.

For all the foregoing reasons and the specialist nature of these systems, it is not considered appropriate to give worked examples from Section X. However, should it be deemed necessary to bill such work, the Rules in Section X are very straightforward.

9 Measurement of Services in Existing Buildings

Introduction

The relevant importance of this topic may be judged by the fact that about 50 per cent of the income generated by the construction industry is attributable to maintenance, improvement, refurbishment and/or extensions of existing properties. This proportion can be even higher during periods of economic recession when new developments are often deferred while essential maintenance and adaptations still proceed.

Work in existing buildings has an additional cost significance to that of any similar work in a new build context. These cost differences can be explained by some or all of the following factors:

(1) Access problems for labour and materials occur within existing structures. These may affect entry to and exit from the point of work in the building and also awkward working conditions arising from the location of existing services, for instance in ducts, cupboards, or behind other services.
(2) Programming of the work is frequently more critical within an existing building because of client requirements and/or continued occupation of the property.
(3) Existing building services often have to be maintained while alterations are carried out, with consequent delays and possible out-of-hours working to suit the client.
(4) Higher skill levels are required by operatives involved in altering existing building services. Also greater care in working is required to prevent damage to surrounding property.
(5) Alteration work is often done in small amounts, which is less efficient to execute.

It will be evident, therefore, that such work should not be combined with similar work in a new build situation. In recognition of the cost significance of work within existing buildings, SMM7 requires all such work to be 'so described', which in effect means kept separate. The relevant rules are General Rules 7.1(a) and 13. The former deals generally with Work of special types, and directs readers to General Rule 13 of which Rule 13.1 defines 'Work to existing buildings' as 'work on, or in, or immediately under work existing before the current project'. This definition recognises that existing work could be very new but has been (or will be) executed prior to the current contract.

Measurement Rules

The general requirement referred to above to keep work to existing buildings separate can be applied in two ways:

(1) Individual items within the body of a bill of quantities labelled 'work to existing building' as appropriate. A typical example could be a new electric main supply cable starting at an existing distribution board within the existing building and led to a new control switch within a new extension. In such a case the cable is likely to be in one continuous length but would necessarily be billed as two items – one item for the normal new work installed in the new extension and one item for 'ditto to existing building'.
(2) Creation of a separate bill or separate section within a bill with an appropriate explanatory heading to include all the items representing work to the existing building. This approach would suit the situation where there is a considerable amount of work of this nature to be billed.

Most of the work to existing buildings comprises items which can equally apply to new works and, apart from the requirements mentioned above, are measured under the normal rules within SMM7. However, there are specific items of work which only apply to installations within existing buildings; for example, breaking into existing pipes, jointing new pipes to existing, cables drawn into existing conduits, and lifting and replacing floor boards for pipes or cables in existing buildings. Such work is covered by the specific SMM7 Rules relating to building services within existing buildings which are found on page 173 covering Sections R10–13 Drainage and Y Mechanical services; page 175 covering Section Y Electrical services; and page 130 covering Sections P30–31 Trenches, holes supports etc. for services. Because of the nature of such work, the majority of these items tend to be enumerated.

Stripping out whole installations may be measured as Y6 page 173 for mechanical, or Y9 page 175 for electrical, or optionally under Section C Spot Items rule C20.2 with reference to C20/M2.

Special Locations of Work

SMM7 rules for mechanical services and, to a lesser extent, for electrical services require that certain special locations are so described and identified separately. Examples are: pipes in ducts, trenches, chases, floor screeds, *in situ* concrete; mechanical work in plant rooms; cables laid in trenches; and conduit in chases. These special locations, should they also occur in existing buildings, would merit the appropriate additional description to the items. The additional costs associated with location are still very relevant within existing buildings.

Taking Off Services in Existing Buildings

In professional practice the task of billing work within existing property requires not only good-quality drawings and specifications, but also a carefully structured site visit by the surveyor to take notes of all the relevant details of the existing services and fabric. Inevitably, drawings cannot show all the requirements and nothing can replace the knowledge gained from site visits.

The worked examples in this chapter cannot, of course, benefit from such a site visit, but readers may find that visiting a similar property and visualising the proposed alterations could be helpful.

Builder's Work in Existing Buildings

The topic of builder's work for services installations in new construction is dealt with in Chapter 2 where it is demonstrated that the subject requires careful consideration. The same requirements of care are necessary when considering the implications of builder's work for new services installations within existing buildings. There are several possibilities which can occur:

(1) New services installation within an existing building which is otherwise not altered in any significant way.

This is the simplest situation as the structure and finishes all pre-exist the services work. The SMM7 rules are P30/31. 32–35. Holes for services are referred in P30/31 Rule M13 to the 'appropriate Work Sections' which are P30/31.20 or Additional rules for Work Groups H, J, K, L and M (SMM7 page 171) Rule 6 – both sets of rules being identical.

(2) New services installation within an existing building which is additionally altered and refurbished.

This introduces the complication of old and new structures and finishes within one location, and the measurement of builder's work should reflect the specific requirements of any particular contract. For instance, if all the plasterwork was being stripped prior to the services being installed then 'making good plaster' would not be required. The SMM7 rules which apply would be as in (1) above for the existing affected structure and finishes, and would be as described in Chapter 2 for new structure and finishes with the added qualification of General Rule 13.

(3) New services installation within an existing building which is additionally altered and extended.

All work outside the outer face of the existing building within the new extension would be wholly new work and treated as such, while work within the existing building would be treated as in (2) above.

WORKED EXAMPLES

Readers should refer to SMM7 and the Measurement Code while following these examples, which will be cross-referenced where appropriate.

These worked examples are set out in draft bill format which clarifies not only the SMM7 measurement requirements but also the various ancillary matters, such as testing and sundries that may not appear clearly in a basic take-off.

Worked Example 9/1

This example demonstrates the application of the Rules of SMM7 to a typical alteration project involving the conversion of a store into a toilet within an existing building. The work includes drainage above and below ground, plumbing, heating and ventilating installations.

Worked Example 9/2

This example demonstrates the application of the Rules of SMM7 to the alterations to a large heating installation of an existing school in order to accommodate an extension to the system serving a new laboratory. (The new heating installation within the laboratory is detailed in Chapter 5.)

Worked Example 9/1: Commercial Building – Services Alterations

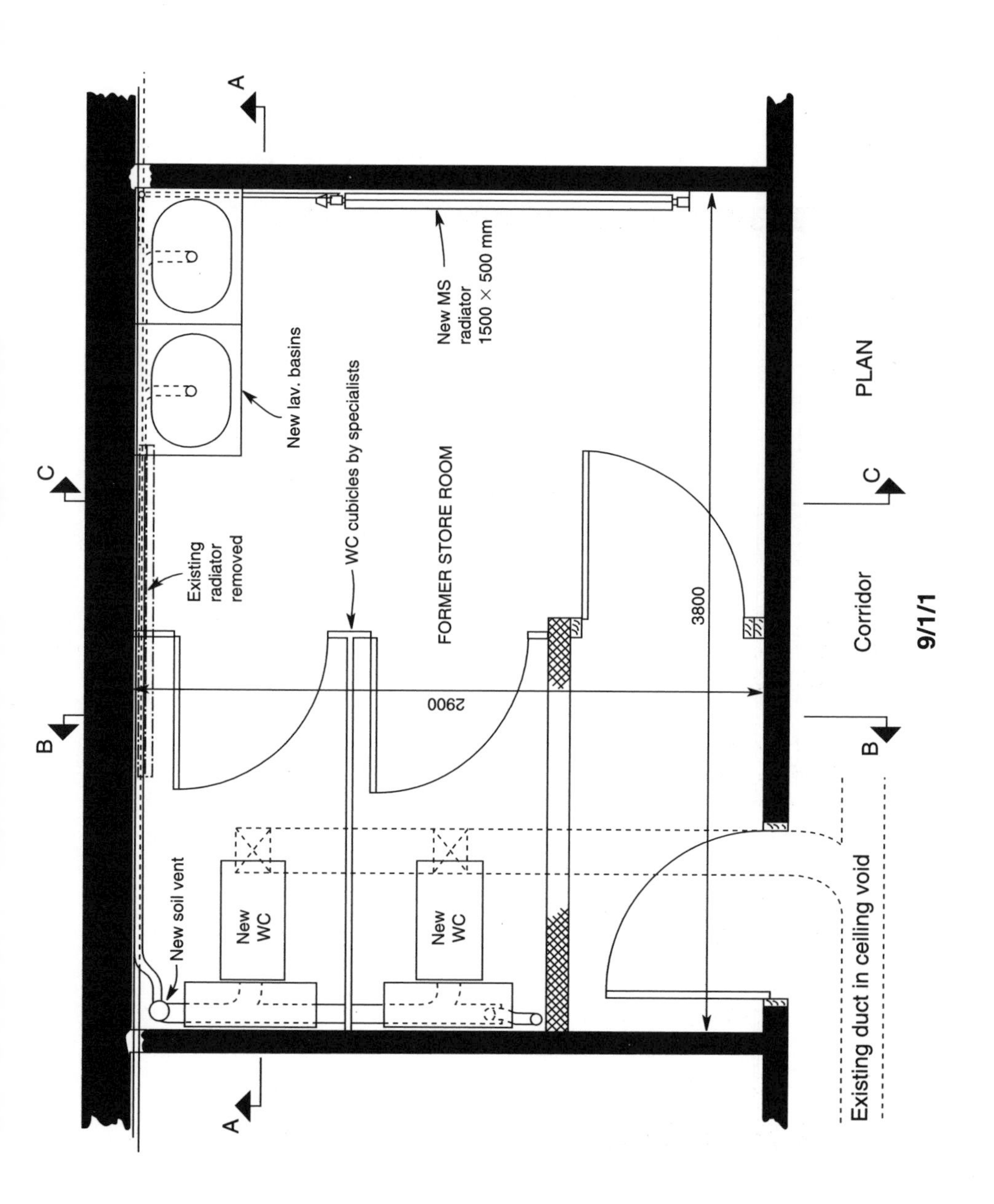

Worked Example 9/1: Commercial Building – Services Alterations

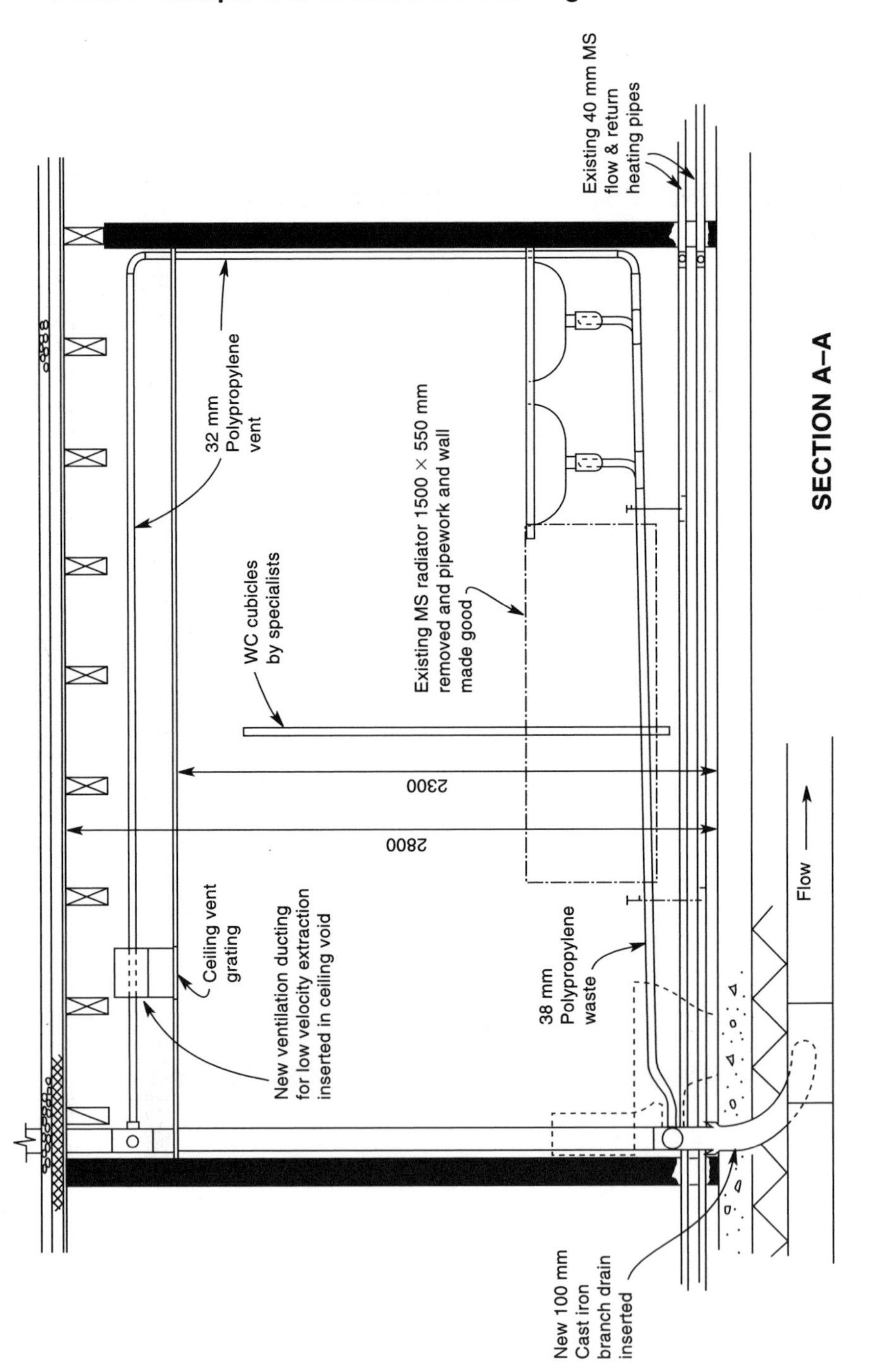

SECTION A–A

9/1/2

Worked Example 9/1: Commercial Building – Services Alterations

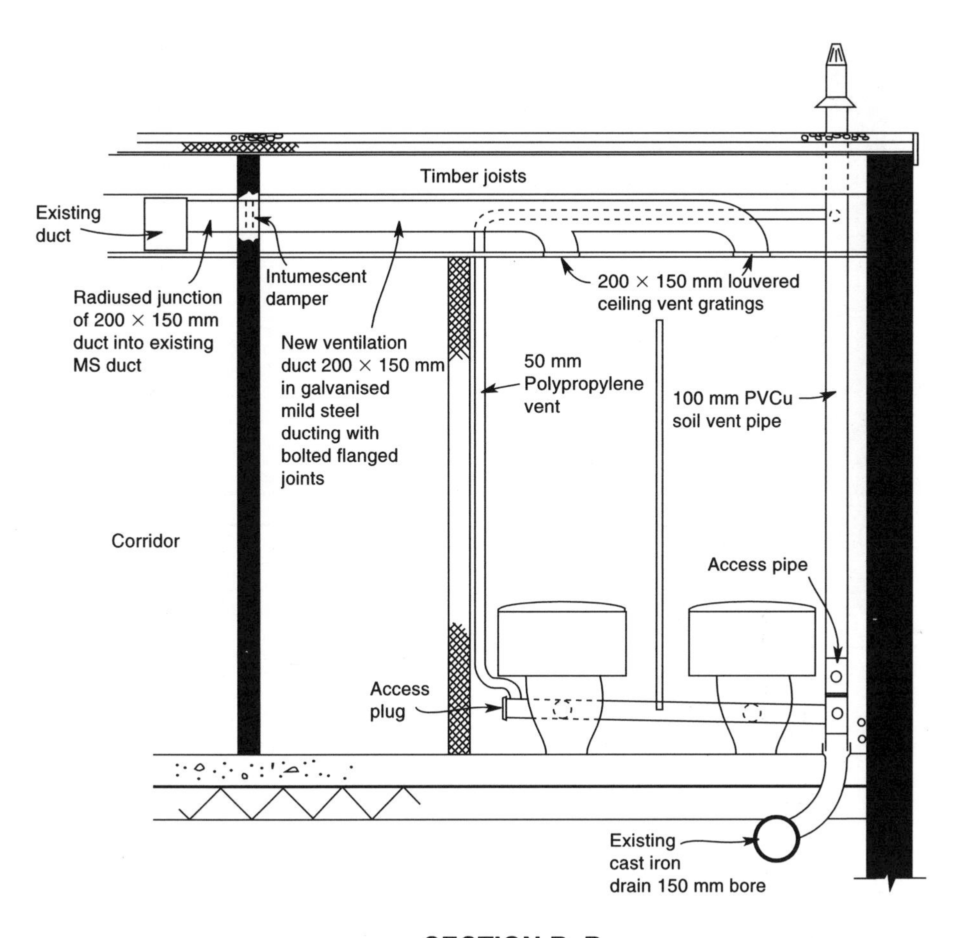

SECTION B–B

9/1/3

Worked Example 9/1: Commercial Building – Services Alterations

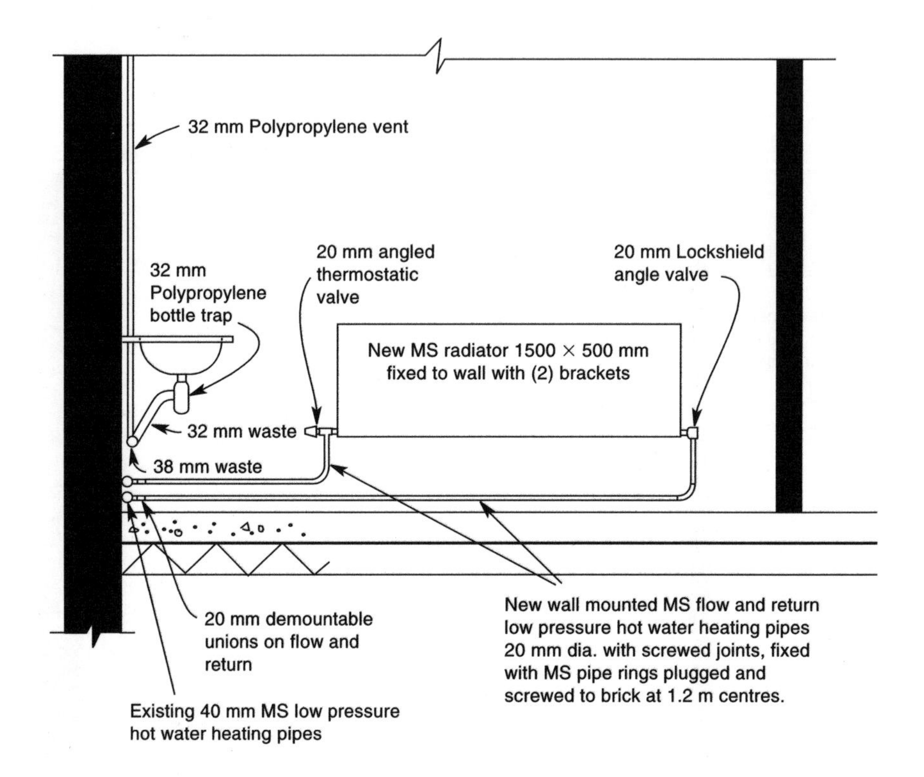

SECTION C–C

9/1/4

Worked Example 9/1: Commercial Building – Services Alterations

NOTES

Connection to existing drain:
Concrete floor broken up to expose 150 mm bore cast iron drain, new 150 × 100 mm branch inserted and 100 mm cast iron branch drain to floor level. Concrete floor made good.

Existing floor:
150 mm reinforced concrete slab on heavy gauge polythene d.p.c. on blinded hardcore 150 mm thick.

Existing walls:
Generally common brickwork with plaster finish.

Existing roof:
Timber joists supporting 18 mm plywood decking, 3 layer built-up felt roofing with 'inverted insulation' comprising 50 mm plastic board insulation topped with 50 mm thick ballast.

Existing ceiling:
Proprietary suspended system comprising aluminium grid suspended from timber joisting with demountable gypsum based plain panels.

Existing ventilation ducting:
Low velocity extraction ducting in corridor ceiling void of galvanised mild steel with bolted flanged joints.

PVCu soil piping:
Nominal 100 mm bore with socketed joints, fixed with GMS pipe clips plugged and screwed to brick.

Polypropylene waste/vent:
To have push-fit 'O' ring couplers and fittings, fixed with plastic pipe clips plugged and screwed to brick.

Radiator:
'Heatwell' Enamel finish mild steel radiator.

WCs:
RICS Sanitary Ware close coupled wash down suites, comprising vitreous china pan and cistern, model No. B/51 complete.

Lavatory basins:
RICS Sanitary Ware vitreous china 600 × 480 mm, model No. C/21 complete with CI wall brackets, CP waste and CP taps type C/42.

9/1/5

WORKED EXAMPLE 9/1				SHEET NR 1
Commentary	Item Nr	Description	Unit	Qty
Project title – assists to reinforce the fact that this bill is entirely concerned with work in existing building as General Rule 13. Sequential order of bills.		*ALTERATIONS TO EXISTING COMMERCIAL BUILDING* *BILL NR 8*		
Non Common Arrangement work heading to include all relevant trades – each of which comprises quantities which are too small to merit separate bills.		*BUILDING SERVICES*		
This note covers the requirements of Rule P1 of various sections & Rule P2 of section P31. Placing first avoids repetition in each subsection of the bill. Clarification that this bill is solely concerned with alterations within existing buildings.		***Information provided*** The following represents the alterations to an existing store room within the existing commercial building to provide new toilet accommodation, comprising soil, waste, drainage, sanitary appliances, hot and cold service, heating and ventilating installations. All as detailed on Plans 9/1/1–5.		
Heading adapted to suit the specific situation. General Rule 13 – stated in each trade heading in case bill split up to seek subcontractor quotes.		***R12 Drainage Below Ground Floor – Work to Existing Building***		
Additional rules SMM7 page 173. Y.1.1.1 Y3.1	1/1	Breaking into existing 150 mm diameter CI drain and insert new 150 × 150 × 100 mm CI drain branch with bolted collar joints, include isolating existing pipe and preparing 2 ends of existing pipe for new; complete with new 100 mm diameter CI drain bend about 600 mm long taking new drain to GFL; in former store (holing concrete floor and making good measured elsewhere) Former store room		item
		R11 Foul Drainage Above Ground – Work to Existing Building		
R11.1		<u>*Grey PVCu soil pipes with socketed joints, jointed with push fit seal*</u>		
R11.1.1.1.1	1/2	Pipes, straight, 110 mm diameter, with GMS bolted pipe clips at not exceeding 2.00 m centres rustproof screwed and plugged to masonry	m	5
Branches included as 'extra over'.		Full ht. 3.20 Branch 1.60 WC branches 2/0.20		
R11.2		*Items extra over 110 mm pipe*		
R11.2.4.5	1/3	Branches	nr	3
		Main stack 1 WC branches 2		

WORKED EXAMPLE 9/1			SHEET NR 2	
Commentary	Item Nr	Description	Unit	Qty
Extra over (cont.)		Extra over 110 mm pipe (cont.):		
	2/1	Stop end with screwed access plug and 50 mm boss	nr	1
		Branch 1		
R11.2.4.5.1	2/2	Access pipe with screwed door	nr	1
		Main stack 1		
R11.2.2.1.1 100 mm CI drain socket larger than 110 mm PVCu pipe.	2/3	Special connection to 100 mm diam. CI drain pipe with plastic connector set in CI socket with mastic	nr	1
		Main stack 1		
R11.2.2.1.1	2/4	Special connections to spigot of WC with flexible plastic connector	nr	2
		Branch 2		
Line to cancel 'extra over'.		- - - - - - - - - - - - - -		
R11.5.1.1 R11/C7 Boss connections are full value and are deemed to include perforating the pipe	2/5	Boss connection utilising cast on boss of previously measured fitting complete with solvent welded adaptor for 'O' ring 50 mm polypropylene pipe	nr	1
		Branch 1		
Alternatively these connections could be formed by using a special fitting with cast on bosses – requiring item plus items similar to 2/5 above .	2/6	Boss connection solvent welded to pipe with 'O' ring connection for polypropylene pipe: 32 mm boss	nr	1
		Main stack 1		
	2/7	50 mm boss	nr	1
		Main stack 1		
R11.6.4.1 Ancillaries are full value and therefore excluded from pipe length measured in item 1/2. R11/C8 Cutting and jointing to pipe deemed included.	2/8	Pipework ancillaries, 110 mm terminal vent cowl	nr	1
		Roof 1		
R11.6.5.1.1	2/9	Pipework ancillaries, 110 mm plastic roof flashing	nr	1
		Roof 1		

WORKED EXAMPLE 9/1			SHEET NR 3	
Commentary	Item Nr	Description	Unit	Qty
		<u>*Polypropylene waste/vent pipes with push fit 'O' ring straight couplings*</u>		
R11.1.1.1.1 Different diameters in sub-items.		Pipes, straight, with plastic pipe clips at 0.60 m centres brass screwed and plugged to masonry		
	3/1	32 mm diameter	m	7
		Basin vent vert. 2.50 Ditto in ceiling 3.65 Branches to basins 2/0.35		
	3/2	38 mm diameter	m	3
		Stack-basins 3.45		
	3/3	50 mm diameter	m	4
		Soil vent vert. 2.45 Ditto in ceiling 1.65		
R11.2.3.*		*Items extra over polypropylene pipes in which they occur*		
Items indented as extra over and additionally indented for each size.		Fittings with two ends;		
	3/4	32 mm diameter	nr	6
		Obtuse bent couplings Basins 2 Vent into stack 2 Bent couplings 2		
	3/5	38 mm diameter	nr	2
		Obtuse bent couplings Waste into stack 2		
	3/6	50 mm diameter	nr	5
		Obtuse bent couplings Vent into stack 2 Bent couplings 3		
	3/7	Fittings with three ends; 38 mm diameter	nr	2
		Swept tees to basins 2 - - - - - - - - - - - - - -		
Line to cancel 'extra over'. R11.6.8.1 Also incorporates R11/D2 as trap is jointed to a sanitary appliance.	3/8	Pipework ancillaries, 32 mm diam. plastic bottle traps with 75 mm seal and screw connection to basin waste	nr	2
		Basins 2		

WORKED EXAMPLE 9/1			SHEET NR 4	
Commentary	Item Nr	Description	Unit	Qty
		N13 Sanitary Appliances/Fittings – Work to Existing Building		
		NOTE:		
Note 1 clarifies N13/C1 as discussed in Chapter 3.		All joints and connections between supply, overflow, waste or soil pipes and the undernoted sanitary appliances/fittings have been measured with the appropriate pipework.		
N13.4.1.1.6 – fully specified items – for 'fix only' approach see Chap. 3.	4/1	Low level close coupled washdown white WC suites comprising vitreous china 'P' trap WC and cistern with black plastic seat and lid, fixed to concrete floor and masonry wall; RICS Sanitary Ware model nr B/51	nr	2
		Former store 2		
	4/2	Lavatory basins 600 × 480 mm comprising white vitreous china basin with CI wall brackets, CP plug and chain waste and CP taps type C/42, fixed to masonry wall; RICS Sanitary Ware model nr C/21	nr	2
No overflows shown on drawing but assume that they will be required. Overflow pipes are not mentioned in SMM7 but logical to measure under section Y10 and immediately after appliances items.		Former store 2		
		Y10 White polythene overflow pipework with solvent welded couplings		
Y10.1.1.1	4/3	Pipes, straight, 20 mm diameter, with plastic pipe clips at not exceeding 0.60 m centres plugged and screwed to masonry	m	2
Assume overflows taken through adjacent outer wall.		2 WCs horizontal 1.55 Ditto vert. 2/0.15		
Y10.2.*		*Items extra over 20 mm pipe*		
Y10.2.3.3 Pipe fittings not exceeding 65 mm diam. simplified measurement	4/4	Fitting with two ends	nr	1
		Bent coupling 1		
Y10.2.3.4	4/5	Fitting with three ends	nr	1
		Branch 1		
Y10.2.1 labour on pipe similar in principle to made bends.	4/6	Neat bevelled cut drip end	nr	1
		Outside 1		
Y10.2.2.1	4/7	Special connections to male boss of WC cistern with female plastic connector	nr	2
		2 WCs 2		

WORKED EXAMPLE 9/1			SHEET NR 5	
Commentary	**Item Nr**	**Description**	**Unit**	**Qty**
Heading from Appendix B with addition of General Rule 13.		***S12 Hot and Cold Water Installation (Small scale) – Work to Existing Building***		
Essential heading to indicate rules adopted for measurements following. Table X – normal quality copper tube.		***Y10/11 Pipelines and Pipeline Ancillaries*** *Copper tubing to BS 2871: Part 1 – Table X*		
Here would follow the detailed measurement of the hot & cold supply pipework which is not shown on the drawings. This measurement would follow the style and format of the equivalent items from the hot & cold installation detailed in the worked example in Chapter 4.				
Heading from Appendix B with addition of General Rule 13.		***T32 Low Temperature Hot Water Heating (Small scale) – Work to Existing Building***		
Additional rules page 173 – heading to indicate rules adopted for measurements following.		***Y Mechanical Services – Work to Existing Building***		
Y.5.1.*.3 Locational information required. Optional nr or item.	5/1	Stripping out part of existing heating system in former store comprising: MS radiator 1500 × 550 mm complete with valves, wall brackets and 2 short lengths of 20 mm F&R pipes; blank off 2 existing tees include isolating and draining down existing system as necessary Former store	item	
Y.1.1.1.3 + 4 Y.3.1 incorporated. Should be measured by 'item' but more effective to enumerate here.	5/2	Breaking into existing 40 mm MS pipes and inserting new 40 × 40 × 20 mm MS tee include isolating and draining down existing pipe, preparing ends of existing and jointing new to old; in former store F&R for new rad. 2	nr	2
Essential heading to indicate rules adopted for measurements following.		***Y10/11 Pipelines and Pipeline Ancillaries*** *Mild steel screwed pipework to BS 1387, table 4, medium weight, jointed with plain sockets with PTFE compound*		
Y10.1.1.1.1	5/3	Pipes, straight, fixed to masonry background with MS pipe rings plugged and screwed at centres as specified; 20 mm diameter Flow 1.25 Return 1.80	m	3

WORKED EXAMPLE 9/1			SHEET NR 6	
Commentary	Item Nr	Description	Unit	Qty
Y10.2		*Items extra over 20 mm pipe*		
Y10.2.1 Items indented once for 'extra over'. No 20 mm bends shown on plans but may occur in reality because of possible obstructions at installation – often taken to have rate in bill.	6/1	Made bends (Say) 2	nr	2
Not in SMM but justified as more labour content than 1 bend but less than 2 bends.	6/2	Made offsets Rise to rad F&R 2	nr	2
Y10.2.3.3.2 Simpler approach to measurement of fittings on pipes not exceeding 65 mm diameter.	6/3	Fittings, MS screwed pattern, two ends Bent couplings F&R 2	nr	2
Y10.2.2.1 – Rule D2 – joints which differ from those in running length.	6/4	Special joints, MS screwed demountable unions F&R 2	nr	2
		Y20/25 General Pipeline Equipment		
Y22.1.1.1.1 + 6.	6/5	Heatwell enamelled steel panel radiator, with (1) 20 mm brass aircock, fixed with concealed MS brackets plugged and screwed to masonry background; 1.12 m^2 heating surface, 1500 × 500 mm Former store 1	nr	1
Y22.2.1.1	6/6	Ancillaries for equipment, thermostatic radiator control angle valve 20 mm diameter once jointed to MS screwed pipe and once jointed to radiator with brass disconnecting union Rad 1	nr	1
	6/7	Ancillaries for equipment, brass radiator lockshield angle valves, 20 mm diameter jointed as last item Rad 1	nr	1
Alternative Measurement Although not strictly in accordance with SMM7 it would be possible to bill the installation of the new radiator and associated pipes and fittings as one detailed enumerated item covering items 5/3–6/7. This could be priced quite successfully by experienced contractors. The item would be of manageable length if the various components were specified fully elsewhere.	Option	Heatwell enamelled steel panel radiator, 1500 × 500 mm with aircock, brackets fixed to masonry, thermostatic and lockshield angle valves all as specification clauses nrs T32/11 & 12. Include connecting to new tees measured elsewhere with about 3 m of 20 mm MS F&R piping fixed with pipe rings to masonry complete with fittings and demountable unions all as shown on Plans 9/1/1–5	nr	1

WORKED EXAMPLE 9/1			SHEET NR 7	
Commentary	Item Nr	Description	Unit	Qty
Heading from Appendix B plus General Rule 13.		***U12 Toilet Extract – Work to Existing Building***		
Locational information is required by certain SMM7 rules but this note clarifies the specific difficulties which could affect the cost of executing this particular installation.		*Note*: All the following ventilation work is located in the existing suspended ceiling of the former store and adjacent corridor. A provisional sum is included elsewhere in the bills to cover the removal, adaptation and reinstatement of the suspended ceiling by specialists to facilitate work by ventilation engineer.		
Additional Rules SMM7 page 173 – Heading to indicate rules adopted for measurements following.		***Y Mechanical Services – Work to Existing Building***		
Y.2.1.1.4 (location required). Y4.1 incorporated. Alternatively Rule Y30.5.1.1.3 could be used. This rule is virtually a repeat of Additional Rule Y2 and appears to be an oversight in SMM7.	7/1	Breaking into existing 200 × 250 mm GMS duct and inserting new radiused junction in 20 gauge GMS with bolted flanged joints to suit new 200 × 150 mm branch duct include preparing ends of existing and jointing new to old; in corridor ceiling outside former store	item	
Essential heading to indicate rules adopted for measurements following.		***Y30/31 Air Ductlines and Air Ductline Ancillaries***		
Y30.1.*.1		GMS 20 gauge ducting, jointed with bolted flanges, supported on GMS straps at centres as specified screwed to timber, suspended immediately below flat roof joists		
Y30.1.1.1.1	7/2	Ducting, straight, rectangular 200 × 150 mm	m	3
Rule Y30–M3		Full length 2.75		
		Branch outlet 0.10		
Inserted branch – item 7/1 is not part of this item.		Deduct branch from item 7/1 Dt.0.15		
Y.30.2.***		*Items extra over 200 × 150 duct*		
Y30.2.3.1	7/3	Fitting, radiused right-angle bend	nr	1
		End of duct 1		
	7/4	Fitting, radiused junction branch	nr	1
		WC cubicle 1		
Y31.4.1.1		*Ancillaries to 200 × 150 duct*		
Measurement Code page 42 for definition of ancillaries.	7/5	Louvered ceiling vent grilles 200 × 150 mm	nr	2
		WC cubicles 2		
	7/6	Intumescent damper as specification clause U12/22	nr	1
		Fire stop partition 1		

WORKED EXAMPLE 9/1		SHEET NR 8		
Commentary	Item Nr	Description	Unit	Qty
		Y51–59 Testing and Sundries – Work to Existing Building		
Y59.1.1 General Rule 13	8/1	Marking positions of holes, mortices and chases in the existing structure for the whole of the foregoing building services installation comprising soil, waste, drainage, sanitary appliances, hot and cold service, heating and ventilating installations	item	
Y51.4.1	8/2	Testing and commissioning the whole of the foregoing building services installation as comprised in previous item all as specification clauses Y51/4–12	item	
P31 Rule M2 – although SMM7 seems to require separate builder's work for various services trades this approach is justifiable in this case.		***P31 Holes, Chases, for Building Services Installations – Work to Existing Building***		
P31.32 Cutting holes in existing buildings – see text of this chapter. Pipe sleeve supply assumed in hot and cold service bill. Assume feeds from corridor ceiling to toilet ceiling.	8/3	Cutting holes for pipes not exceeding 55 mm nominal size through existing half brick wall in ceiling void, take delivery of and build in pipe sleeve and making good	nr	2
		H&C to new toilet 2		
P31.32 as P20.1.1.1	8/4	Cutting hole for rectangular duct girth not exceeding 1.00 m through existing half brick wall in ceiling void, and making good	nr	1
		Ventilation 1		
P31.32 + M13 Additional rules SMM7, page 171, Rule 6.2.3.4	8/5	Cutting hole for new drain connection to existing drain exceeding 110 mm nominal size in existing concrete floor comprising 150 mm thick reinforced concrete bed on heavy gauge polythene DPM on blinded hardcore in former store, and making good incorporating cold emulsion brush applied repair to DPM	nr	1
		Hole for item 1/1 1		
P31.32 + M13 Additional rules SMM7, page 171, Rule 6.2.3.4	8/6	Cutting hole for pipe 55–110 mm nominal size through existing roof comprising timber joists supporting 18 mm plywood deck with built up roofing and inverted insulation with 50 mm ballast, and making good (suspended ceiling removed and reinstated by others)	nr	1
		Soil pipe 1		

Worked Example 9/2: Extension of School Heating Installation (to be read in conjunction with Drawings for Worked Example 5.2)

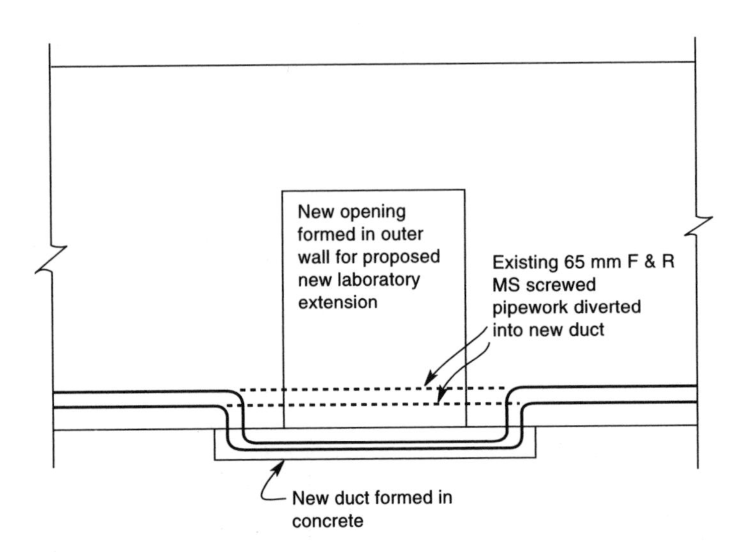

INTERNAL ELEVATION – EXISTING CORRIDOR

9/2/1

Worked Example 9/2: Extension of School Heating Installation (to be read in conjunction with Drawings for Worked Example 5.2)

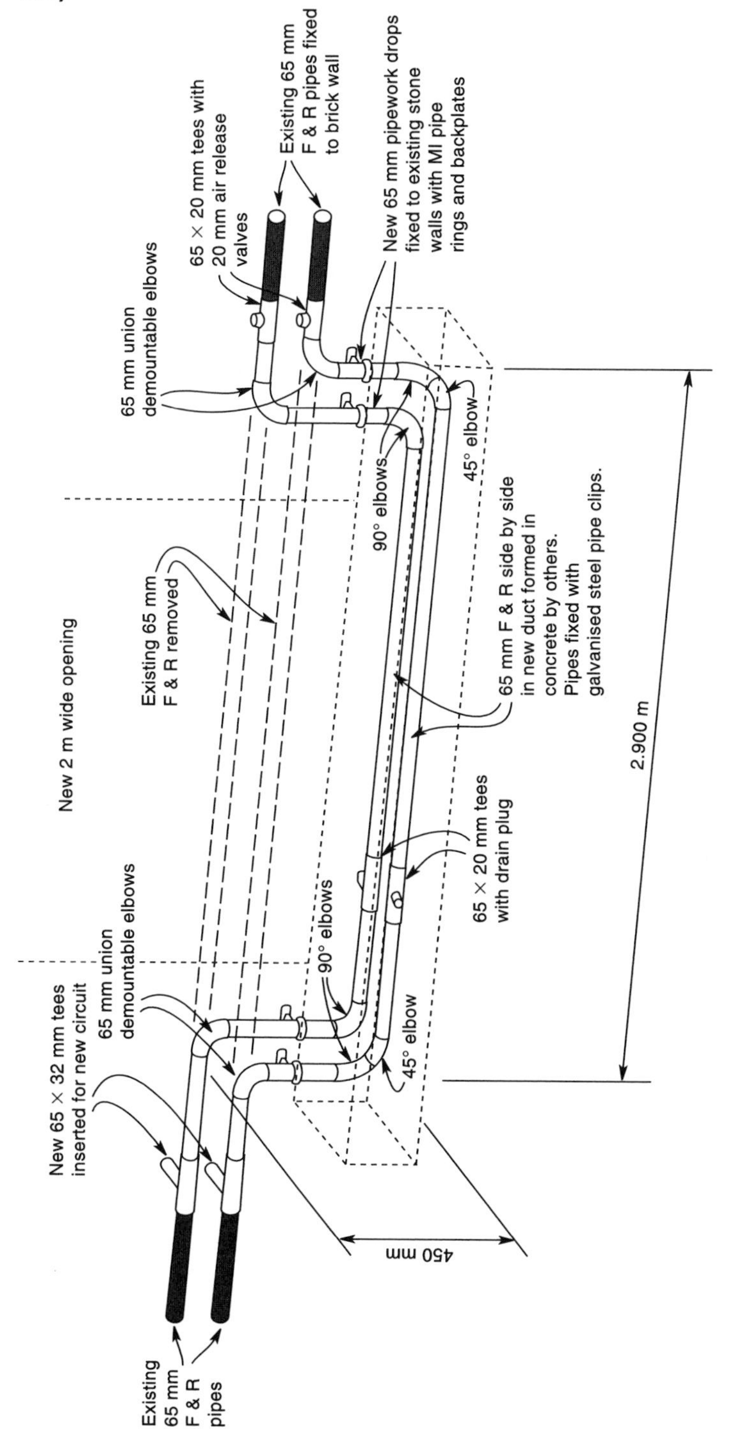

DETAIL OF DIVERSION OF EXISTING HEATING PIPES AT NEW OPENING

9/2/2

WORKED EXAMPLE 9/2			SHEET NR 1	
Commentary	Item Nr	Description	Unit	Qty
Project title – assists to reinforce the fact that this bill is entirely concerned with work in existing building as General Rule 13. Sequential order of bills.		*LABORATORY EXTENSION – ALTERATIONS TO EXISTING SCHOOL* *BILL NR 11*		
Common Arrangement of Work Section heading		*T – MECHANICAL HEATING SYSTEM*		
General Rules 7.1(a) & 13.		***Work to Existing Building***		
		Information provided		
This note covers the requirements of Rule P1 of sections Y10–59 & Rule P2 of section P31. Placing first avoids repetition in each subsection of the bill. Clarification that this bill is solely concerned with alterations within existing buildings.		The following represents the alterations to the existing heating system within the existing school to provide diversions and connections for the new laboratory extension heating installation separately measured in Bill Nr 10. All as detailed on Plans 9/2/1 & 2.		
Heading from Appendix B.		***T31 Low Temperature Hot Water Heating***		
Essential heading to indicate rules adopted for measurements following.		***Y10/11 Pipelines and Pipeline Ancillaries***		
		Mild steel screwed pipework to BS 1387, table 4, medium weight, jointed with plain sockets with PTFE compound		
Additional Rule Y.1.1.1.3. Optionally could be Y.5.1.	1/1	Break into existing 65 mm MS flow and return pipes in corridor and remove sections 3.50 m long from both pipes to clear wall for formation of new opening, include isolating and draining down both pipes	Item	
Additional Rule Y.3.1.*.1 – care to include actual jointing which is not an option in Additional Rule Y.1.1.1.*.	1/2	Jointing new screwed pipe fittings (fittings measured later) to existing 65 mm MS pipes include preparing pipe ends	nr	4
		All ends in item 1/1 = 2 × 2		
Direct take off – abstract unnecessary. Y10.1.1.1.1	1/3	Pipes, straight, fixed to masonry background with MS pipe rings plugged and screwed at centres as specified; 65 mm diameter	m	3
		Left of opening F 0.90 R 0.70 Right of opening F 0.70 R 0.50		
Special location Y10.1.**.2	1/4	Pipes, straight, fixed to masonry background with GMS pipe clips plugged and screwed at centres as specified; in duct; 65 mm diameter	m	7
		New duct in corridor F 3.40 R 3.20		

WORKED EXAMPLE 9/2			SHEET NR 2	
Commentary	Item Nr	Description	Unit	Qty
Pipe items (cont.)		*Mild steel screwed pipework (cont.)*		
Two short lengths of pipe which are part of the F&R continuing into the new extension.	2/1	Pipes, straight, not fixed; 32 mm diameter	m	1
		Through old wall 2/0.50		
Y10.2		*Items extra over MS pipes in which they occur*		
Y10.2.3.3.2 Simpler approach to measurement of fittings on pipes not exceeding 65 mm diameter.	2/2	Fittings, MS screwed pattern, two ends; 65 mm pipe	nr	6
		Elbows 90 4 45 2		
Y10.2.3.4.2	2/3	Fittings, MS screwed pattern, three ends; 65 mm pipe	nr	6
		Tees 6		
Y10.2.2.1 – Rule D2 – joints which differ from those in running length.	2/4	Special joints, MS screwed demountable elbows; 65 mm pipe	nr	4
Broken line to end 'extra over'.		Drops to duct 4 - - - - - - - - - - - - - -		
Y11.8 + Measurement Code Full value. Method of jointing stated.		*Pipework Ancillaries*		
	2/5	Brass air release valves screw jointed to MS pipe; 20 mm diameter	nr	2
		Tees F&R 2		
	2/6	Brass drain down plugs screw jointed to MS pipe; 20 mm diameter	nr	2
		Tees in duct F&R 2		
Y11.11.2.1.2	2/7	Pipe sleeves through walls, length not exceeding 600 mm in MS for 32 mm pipe, handed to builder for fixing	nr	2
		450 mm existing wall 2		
Specification states all pipes in ducts to be insulated. Y50.1.1.1		***Y50 Thermal Insulation***		
		Plastic faced glass fibre sectional insulation minimum 15 mm thick, secured with waterproof self-adhesive tape		
	2/8	Pipelines, 65 mm diameter	m	7
Qty. item nr 1/4 Y50.2.1.1		Qty. pipe in duct 6.60		
	2/9	Extra over pipeline insulation for working around ancillaries – drain down cocks	nr	2
		Duct 2		

WORKED EXAMPLE 9/2		SHEET NR 3		
Commentary	Item Nr	Description	Unit	Qty
		Y51–59 Testing and Sundries		
Y54.1.1 General Rule 13	3/1	Marking positions of holes, mortices and chases in the existing structure for the whole of the foregoing hot water heating system	item	
Y51.4.1	3/2	Testing and commissioning the whole of the foregoing hot water heating system as per specification clause Y51/4	item	
P31 Rule M2 P31.32 Cutting holes and sinkings for services installations in existing buildings – see text of this chapter.		***P31 Holes, Chases, for Hot water heating system***		
Pipe sleeve supply in item 2/7.	3/3	Cutting holes for pipes not exceeding 55 mm nominal size through existing stone wall 450 mm thick, take delivery of and build in pipe sleeve and making good	nr	2
		F&R to new extension 2		
P31.32	3/4	Cutting sinking to form pipe duct in existing concrete floor approximately 250 mm wide × 150 mm deep x 3200 mm long; install MS angle frame surround with 3 mm thick MS chequer plate flush cover in two pieces each twice notched for 32 mm pipes; and making good	nr	1
		Existing corridor 1		

10 Trade Format Bills

Introduction

Trade format of bills of quantities can be defined as bills arranged so that all work belonging to a traditional trade is grouped together under an appropriate heading.

The rather strange fact that 'Plumbing Work' does not feature in the Alphabetical Index of SMM7 has already been raised in Chapter 2. This is explained by the fact that SMM7 has quite deliberately been produced on a work section basis rather than on a trades basis. SMM7 was developed as a result of a drive by construction professionals in the late 1970s and the 1980s to improve the quality and the presentation of contract information. This work was overseen by the 'Co-ordinating Committee for Project Information' who developed the 'Common arrangement of work sections for building works' usually shortened to 'CAWS'. The development of work sections format was in the context of some very large and complex building projects current at that time, when management contractors and project managers were administering contracts comprising 100–200 specialist subcontractors. Much of the work in the construction industry is, however, of more modest value and complexity with about 80 per cent of the annual building work budget being carried out by firms with fewer than 20 employees. These smaller firms normally organise and operate on a traditional trade basis and it is this fact that justifies the contents of this chapter.

Trade Format in Building Services

The Measurement Code comments on the format of bills of quantities on page 7 as follows: 'The format of bills of quantities continues to be a matter for the discretion of the surveyor preparing bills of quantities for a particular project'. This proviso leaves the decision on the arrangement of the bill, rightly, with the person who can make the most appropriate choice in the particular circumstances.

The format of contract documentation produced for any project should be relevant to the value, size and complexity while being essentially geared to the needs of the users of the documentation. In larger works there is every justification for following the CAWS format but for smaller jobs the option to produce bills of quantities on a trade format basis should be considered, particularly where the likely contractors will be small firms organised in traditional trades.

The common trades traditionally involved with building services are:

Plumber
Heating and Ventilating Engineer
Electrician

It is relatively straightforward to produce bills arranged to suit these trades but it is important that the appropriate SMM7 headings are incorporated in the usual way. This requirement permits tendering contractors to be fully aware of the rules under which the work has been measured. The resulting trade format bills, having collected the relevant work into readily accessed locations, simplify the task of main contractors seeking subcontractor quotes from the various building services trade subcontractors. In the post-contract situation, such bills can also simplify the preparation of interim valuations and final accounts.

WORKED EXAMPLE

The worked example to illustrate a trade format bill is based on the PLUMBING WORK of the large house project used in the worked examples in Chapter 3 and Chapter 4 covering Disposal Systems and Service Plumbing Work. This size of project, being a contract for a one-off detached house, is typical of the circumstances where trade format bills could be adopted.

The traditional trade of plumbing encompasses such additional features as sheet metal flashings which are not part of building services but are usually included in a trade bill. Conversely, below ground drainage is likely to be part of the builder trade and is therefore omitted here.

As the services work has already been measured in detail in earlier chapters, only the bill layout with headings is given to illustrate the trade format. The items have been omitted for brevity.

Reference should be made to Chapters 3 and 4 for the drawings and detailed measurements of this worked example.

WORKED EXAMPLE 10/1		SHEET NR 1		
Commentary	Item Nr	Description	Unit	Qty
		DETACHED HOUSE		
Sequential order of trade bills.		*BILL NR 3*		
Non SMM7 trade heading.		*PLUMBING WORK*		
		Information provided		
This note covers the requirements of Rule P1 of sections H71, R10 & 11, N13 and Y10–59. Placing this first removes the need for repetition within the various subsections of the bill.		The following represents the installation of lead flashings, rainwater, foul drainage above ground, sanitary appliances and cold and hot service plumbing as detailed on Plans 3/1/1–5 & 4/1/1 & 4/1/2.		
Roof leadwork traditionally part of plumbing work. Only flashings in this example thus 'coverings' omitted.		***H71 Lead Sheet Flashings***		
		Milled sheet lead to BS 1178, Nr 5 gauge in roof flashings as specification clause H71/3		
		Detailed items		
		R10 Rainwater Pipework/ Gutters		
		Grey PVCu rainwater pipework with socketed joints		
		Detailed items		
		Grey PVCu rainwater gutter with plain ends and union brackets		
		Detailed items		
		R11 Foul Drainage Above Ground		
		Cast iron soil pipes to BS 460, heavy quality coated, with socketed joints, jointed with rope yarn and molten lead, work below ground floor level in substructure		
		Detailed items		
		Grey PVCu soil pipes with socketed joints, jointed with push fit seal		
		Detailed items		
		Polypropylene waste pipes with push fit 'O' ring straight couplings		
		Detailed items		

WORKED EXAMPLE 10/1		SHEET NR 2		
Commentary	Item Nr	Description	Unit	Qty
		N13 Sanitary Appliances/Fittings		
		NOTE:		
Note 1 clarifies N13/C1 as discussed in Chapter 3.		(1) All joints and connections between supply, overflow, waste or soil pipes and the undernoted sanitary appliances/fittings have been measured with the appropriate pipework.		
Sanitary appliances may be billed in a direct manner quoting maker's catalogue references for each item but in practice a provisional sum is often used to permit the final choice of unit to be left to nearer the time of installation. 'Fixing only' is defined in SMM7 Preliminaries/General Conditions – A52 Nominated Suppliers Rule C1 (SMM7 amendment 2 Feb. 1989).		(2) The undernoted sanitary appliances/fittings are included in a Provisional Sum for defined work contained in Section 2 of this bill to cover the supply and delivery to site of these items by a Nominated Supplier.		
		Fixing only the following sanitary appliances/fittings:		
		Detailed items		
		S10 Cold Water Installation – Main Supply – Underground		
		Y10/11 Pipelines and Pipeline Ancillaries		
		Copper tubing to BS 2871: Part 1 – Table Y		
		Detailed items		
		S10 Cold Water Installation – Main Supply – Internal		
		Y10/11 Pipelines and Pipeline Ancillaries		
		Copper tubing to BS 2871: Part 1 – Table X		
		Detailed items		
		Pipework Ancillaries		
		Detailed items		
		Y20/25 General Pipeline Equipment		
		Detailed items		
		Y10 Polypropylene overflow pipework jointed with push fit 'O' ring couplings		
		Detailed items		
		S10 Cold Water Installation – Cold Supply		
		Y10/11 Pipelines and Pipeline Ancillaries		
		Copper tubing to BS 2871: Part 1 – Table X		
		Detailed items		

WORKED EXAMPLE 10/1		SHEET NR 3		
Commentary	Item Nr	Description	Unit	Qty
		S11 Hot Water Installation		
		Y10/11 Pipelines and Pipeline Ancillaries		
		Copper tubing to BS 2871: Part 1 – Table X		
		Detailed items		
		Pipework Ancillaries		
		Detailed items		
		Y20/25 General Pipeline Equipment		
		Detailed items		
		Y50 Thermal Insulation		
		Plastic faced glass fibre sectional insulation minimum 15 mm thick, secured with waterproof adhesive tape		
		Detailed items		
		Insulation packs in preformed foam as specified manufactured by Insulpak plc		
		Detailed items		
		Y51–59 Testing and Sundries for whole of Plumbing Work		
		Detailed items		
		P30/31 Trenches, Holes, Chases for whole of Plumbing Work		
P30 Information Provided not covered at start of this Bill – therefore required here.		***Information Provided:*** for information regarding the nature of excavation work refer to Bill Nr 1 – Builder and Drainage Works.		
		Detailed items		

Bibliography

Royal Institution of Chartered Surveyors and Building Employers Confederation, *Standard Method of Measurement of Building Works: Seventh Edition (SMM7)* (1988)

Royal Institution of Chartered Surveyors and Building Employers Confederation, *A Code of Procedure for Measurement of Building Works (SMM7 Measurement Code)* (1988)

Building Project Information Committee, *Common Arrangement of Work Sections for Building Works* (1987)

Building Project Information Committee, *Project Specification: A Code of Procedure for Building Works* (1987)

Co-ordinating Committee for Project Information, *Co-ordinated Project Information for Building Works: a guide with examples* (1987)

E.F. Curd and C.A. Howard, *Introduction to Building Services*, 2nd edn, Macmillan (1996)

I.H. Seeley, *Advanced Building Measurement*, 2nd edn, Macmillan (1990)

I.H. Seeley, *Building Quantities Explained*, 4th edn, Macmillan (1992)

Index